Curators

*To Chris O'Herlihy,
With best wishes.*

[signature]

Curators

Behind the Scenes of Natural History Museums

Lance Grande

The University of Chicago Press
Chicago and London

The University of Chicago Press, Chicago 60637
The University of Chicago Press, Ltd., London
© 2017 by Lance Grande
All rights reserved. No part of this book may be used or reproduced in any manner whatsoever without written permission, except in the case of brief quotations in critical articles and reviews. For more information, contact the University of Chicago Press, 1427 East 60th Street, Chicago, IL 60637.
Published 2017.
Printed in the United States of America

26 25 24 23 22 21 20 19 18 17   1 2 3 4 5

ISBN-13: 978-0-226-19275-8 (cloth)
ISBN-13: 978-0-226-38943-1 (e-book)
DOI: 10.7208/chicago/9780226389431.001.0001

Library of Congress Cataloging-in-Publication Data
Names: Grande, Lance, author.
Title: Curators: behind the scenes of natural history museums / Lance Grande.
Description: Chicago; London: The University of Chicago Press, 2017. | Includes bibliographical references and index.
Identifiers: LCCN 2016032596| ISBN 9780226192758 (cloth: alk. paper) | ISBN 9780226389431 (e-book)
Subjects: LCSH: Grande, Lance. | Natural history museum curators—Illinois—Chicago—Biography. | Field Museum of Natural History—Biography. | Biologists—Illinois—Chicago—Biography. | Paleontologists—Illinois—Chicago—Biography. | Natural history museum curators. | Curatorship. | Natural history museums.
Classification: LCC QH31 .G67 2017 | DDC 508.092 [B]—dc23 LC record available at https://lccn.loc.gov/2016032596

♾ This paper meets the requirements of ANSI/NISO Z39.48-1992 (Permanence of Paper).

To Dianne,

for your sweet love and support.

To Lauren, Elizabeth, Patrick, and Kevin,

that you each follow your own ambitions

with joy and conviction.

Fulfillment is more about the journey

than the destination.

CONTENTS

Preface: Curators of Natural History and Human Culture ix

1. Moving toward the Life of a Curator *1*
2. Beginning a Curatorial Career *23*
3. Staking Out a Field Site in Wyoming *35*
4. Mexico and the Hotel NSF *65*
5. Willy, Radioactive Rayfins, and the Fish Rodeo *85*
6. A Dino Named SUE *111*
7. Adventures of My Curatorial Colleagues from the Field *157*
8. The Spirit of K-P Schmidt and the Hazards of Herpetology *197*
9. Executive Management *211*
10. Exhibition and the Grainger Hall of Gems *233*

11	Grave Concerns *259*
12	Hunting–and Conserving–Lions *285*
13	Saving the Planet's Ecosystems *307*
14	Where Do We Go from Here? *339*

Acknowledgments 353

Notes, Added Commentary, References, and Figure Credits 355

Index 407

PREFACE

Curators of Natural History and Human Culture

I am a curator: one of the primary research scientists at the Field Museum in Chicago. I've been a curator for more than thirty-three years. I was inspired by curators before me who were influenced by curators before them.

The Field Museum is one of the largest natural history museums in the world. It houses more than 27 million specimens ranging from DNA to dinosaurs.[1] For more than 120 years, curators have assembled this collection to study and document our planet's biology, geology, and human culture. Their research has provided unique insight into the history and diversity of our world. While leading the Research and Collection division of the museum for more than eight years as a senior vice president, I came to realize that few people understood what a natural history museum curator does. That realization was the first impetus for me to write this book.

In major natural history museums of North America, the term *curator* is used for the research scientist whose job is to bring authority and originality to their museum's scientific message. They explore the natural and cultural world through field expeditions; they do original research based on objects of natural or cultural history ("specimens" and "artifacts"); they disseminate knowledge of original scientific discoveries to students, other scientists, and the general public. Curators are passionate about their quest for understanding the Earth and its people. They challenge convention, sometimes even at their own peril, to enable scientific progress.

The origins of curators and museums of cultural and natural history began long ago. The earliest known cultural history museum was established over 2,500 years ago by a Babylonian princess and her father in the ancient Mesopotamian city of Ur (now the Dhi Qar Governorate of Iraq). Princess Ennigaldi was a high priestess of the moon god Nanna, and the daughter of Nabonidus, last king of the Neo-Babylonian Empire. Nabonidus collected antiquities and is the earliest known archaeologist. In 530 BC he influenced his daughter to develop a museum focusing on the cultural history of Mesopotamia. She became the museum's first curator and developed a research program around its collection of artifacts. This was centuries before the study of natural history was established, and her study of culture was primitive by today's standards. All advances in science must be considered within the context of their own time, and the idea of a research facility with an archived collection was a huge step for the sixth century BC.

Ennigaldi's museum was active until around 500 BC when the city of Ur was abandoned due to deteriorating environmental conditions (prolonged drought and changing river conditions). There are no records of what happened to the princess. The museum was lost for thousands of years until it was rediscovered in 1925 by the famous archaeologist Leonard Woolley. While excavating an ancient Babylonian palace, Woolley discovered a large chamber with a curious collection of neatly arranged artifacts ranging in age from 2100 BC to 600 BC. The objects were associated with a series of inscribed clay cylinders representing labels. Woolley soon realized that he had discovered the remains of the world's oldest known museum.

The earliest known center for natural history research goes back about 2,300 years, to the third century BC. The Lyceum, in classical Athens, was a center of scholarly research and learning where Aristotle developed one of the earliest hierarchical classifications of living things. He proposed a *Scala naturae* ("Great Chain of Being," also called "Ladder of Life") to classify plants and animals according to their structure. This classification, while not as sophisticated or as comprehensive as the later classification of Carolus Linnaeus, was one of the first to use the internal structures of organisms. Today Aristotle is considered to be the founder of comparative anatomy and the first genuine natural history scientist. He pioneered the study of zoology and made significant contributions to the studies of geology, botany, physics, and philosophy. Aristotle's work was ahead of its time, and he eventually angered high religious officials, who denounced him in 322 BC for not holding the gods in honor. He fled Athens, fearing for his life, and died later that year of natural causes. Although Aristotle's research was specimen-based, he left no surviving collection of museum specimens. In fact, we have little record of natural history collections made until the sixteenth century.

Through the sixteenth and seventeenth centuries, natural history collections were generally personal accumulations of specimens and artifacts organized by people who made the bulk of their income as physicians, professors, or advisors to royalty or religious authorities. These collections were often associated with small museums that amounted to little more than what were called "cabinets of curiosities" (the word *cabinet* referring to a room rather than a piece of furniture). These displays of eclectic and poorly organized objects were precursors to modern-day natural history museums. The collections sometimes blended fact and fiction, featuring faked mythical creatures (e.g., unicorns, mermaids, dragons, and gryphons) made from parts of real animals stuck together by barber surgeons. (At that time, surgery was the charge of barbers rather than physicians.) The purpose of a cabinet of curiosities was not scientific as we understand science today. Instead, these crude exhibits were meant to be theaters of wonder, propaganda, or even displays of personal wealth and power. Although many of the objects were authentic pieces of nature and antiquity, few of them ever made their way into modern museum collections.

During the eighteenth century, the concept of a museum collection gradually evolved from ephemeral displays of curios to being systematically organized libraries of objects documenting natural history and human culture. As the collections increased in sophistication, they became associated with scientifically minded research curators who developed improved methods of organization. One of history's earliest curators of natural history specimens, whose systematically organized collection still exists today, is also one of history's most important biodiversity specialists: Carolus Linnaeus (1707–1778). Linnaeus is widely regarded as the father of modern taxonomy (the branch of science concerned with classifying, naming, and organizing species). He had a doctorate degree and the equivalent of graduate students studying under him—although in his day obtaining a doctorate often involved a much simpler and shorter process than today, and he referred to his students as his "apostles." Linnaeus left a lasting legacy for all curators and systematic biologists who came after him by creating the hierarchical classification system for all plants and animals that we still use: Species within Genera, Genera within Families, Families within Orders, Orders within Classes, Classes within Phyla, and Phyla within Kingdoms. Just as the Dewey decimal and Library of Congress classifications enable efficient retrieval of a single book volume from large libraries, the Linnaean system allows us to efficiently find information on a single species among the tens of millions that exist in nature. Linnaeus made extensive collections of plants and animals through fieldwork in several different countries, published many books on his research using these collections, and preserved much of his material for future generations. He named and classified over 13,000 species of plants and animals, and left a large collection of specimens for reference purposes. The collection of Linnaeus is one of the oldest natural history collections that is still curated and used today. The Linnean Society of London and the Swedish Museum of

Natural History in Stockholm house much of the collection, and I have used some of these specimens in my own research.

Linnaeus, like most people of his time, started out as a strict creationist, believing that each and every species was independently created by God over the span of six days, as narrated in the Judeo-Christian book of Genesis. Toward the end of his life after decades of research on plant and animal species, he started to interpret certain natural patterns based on shared similarities among groups of species. He hinted at a sort of evolution of some species, although not in the Darwinian sense as we understand it today. He came to the conclusion that within every genus only one species had been divinely created. He proposed that the rest of the species developed later through natural processes (although he guessed that process to be hybridization between the mother and other species). The idea that all species were not independently created by God during the six days of creation evoked strong warnings from both the Catholic and Protestant churches. The Lutheran Archbishop of Uppsala accused him of impiety. In 1758 Pope Clement XIII banned the works of Linnaeus by listing them in the *Index Librorum Prohibitorum*. Linnaeus faced a serious backlash by being ahead of his time and questioning dogmatic authoritarianism based on scientific observation of nature. Curators are often the ones to make fundamental discoveries that change the way we look at life and culture by challenging dogmatic ideas and cherished beliefs. Historically, this has sometimes put them at odds with institutional administrators, senior scientists in their fields of study, religious authorities, or even government officials.[2]

Through the eighteenth and nineteenth centuries, museum collections increased in size and diversity. As they grew, new fields of scientific research developed, such as geology, paleontology, evolutionary biology, and anthropology. Fossils were no longer seen as remnants of mythical creatures or as odd-shaped stone objects of non-organic origin. Scientists began to recognize them as extinct species within Earth's complex history of life. Oddly chipped pieces of stone more than a million years old resembling scraping knives and weapons were now recognized as Stone Age tools made by early hominid species. Countless records of primitive and extinct human cultures were found buried all around the world and collected. The history of how and why the world came to be the way it is was revealing itself.

With the development of modern sciences, museums continued to grow and evolve. In their book *Life on Display: Revolutionizing U.S. Museums of Science and Natural History in the Twentieth Century*, history professors Karen Rader and Victoria Cain recount how in the 1870s and 1880s American natural history museums began as little more than shrines to scientific knowledge and authority whose only defining features were the accumulation and dull display of specimens and artifacts. The main charge of curators was simply to amass collections and to put part of them in exhibit cases for public view. A general lack of inspiration served to stereotype natural history museums as musty, dusty storehouses of stuffed animals and bugs with no popular interest and of little use to society. By the 1890s, leaders of major natural history museums in the United States realized that they needed to improve their declining image, so they developed what was called the "New Museum Idea." This concept proposed that natural history museums should do more than simply amass collections of objects. They could better serve the public first by producing new scientific knowledge, and second by disseminating that knowledge to the general public more effectively.

Within the framework of the New Museum Idea, both goals could be served by giving curators more time for research. Management of exhibit development and design became more of an independent entity within major museums. By the late twentieth century, all of the major museums had created independent exhibitions departments staffed with large numbers of non-curatorial specialists (e.g., a peak of around seventy staff positions at the Field Museum in the 1990s). Full-time curator of exhibits positions were phased out of the scientific departments, and the curator's role in exhibit production became that of an opportunistically chosen content specialist. Incorporation of professional designers into exhibit production gave exhibits a more polished look and made natural history museums more effective in the delivery of educational entertainment. Exhibits better engaged the public in ways that improved science literacy and the understanding of nature and culture. With fewer exhibition responsibilities, curators were able to give greater focus to their research efforts, enhancing their ability to make new and important scientific discoveries. With more time for their research, curators also developed a greater role in higher education, making curatorial positions more akin to university professors.

Curators in today's major natural history museums have specialized roles, both pragmatically and intellectually. They are internationally collaborative yet independently driven. They are expected to leverage their curatorial positions into robust scientific programs. Their job is to make new discoveries through exploration and research, to write grant proposals to help support and build those research programs, and to publish their results in scientific journals, monographs, and books. They focus on subjects ranging from the diversity and evolution of species, to the development of complex human societies, or even to the origin of the solar system. Their research forms the basis for our understanding of the most central issues of our existence: biodiversity, cultural history, and humankind's place in the ever-changing network of life on Earth. Curators often lead a life of adventure, traveling the world to do fieldwork and research. They build permanent collections that function as an empirical library of life and culture for the benefit of human society.

Although collection building is an important part of the curatorial story, this book focuses mainly on the role of the curator as the museum's primary scientist. Who are they? What do they do? Where do they go to do it? At the end of this book, there is a chapter-by-chapter section that includes endnotes and reference sources for further reading for those interested in more detail.

As curators, most of our publications are fairly impersonal technical articles written for an audience of other scientists, university students, and special interests. Those publications typically dwell on our discoveries rather than the extensive work it took to achieve each success. In contrast, this book presents an inside view of the curatorial profession from a very personal context. The common thread I chose for this book involves people and events connected to my own intellectual and professional development—from beginning student to senior curator. Chapters 1–3 focus on people and events that led to my becoming a curator. The first chapter includes a brief technical description of cladistic scientific method. The clash of ideologies during the rise of cladistics showed me the intensity of controversy that can occur within the professional scientific community. Such battles are often forgotten in the shadow of an emerging consensus, but the process of these internal struggles is an important aspect of scientific progress. Being part of that turmoil as a graduate student and early professional made me a much better scientist.

Chapters 4–8 are primarily about curatorial colleagues and events that influenced my development *as* a curator. They broadened my appreciation for curatorial dedication as well as for the diversity of curatorial research and responsibilities. Chapters 9–10 contain experiences that deepened my knowledge of museums, including stories behind iconic specimen acquisitions, administrative leadership, and exhibition development. Chapters 11–13 include evolving issues for curators and natural history museums such as repatriation, collection ethics, and conservation. Lastly, chapter 14 reflects on growing challenges for natural history museums and how the role of curators and museums might evolve to address them.

Writing this book was an emotional experience. Perhaps in some ways my approach makes this less of a scientific book and more of a historical document commemorating museum research curators. So be it.

Opening day of the Field Museum in Chicago, June 2, 1894. The missions of natural history museums and their curators have evolved tremendously over the last century. My personal immersion into natural history began in Minnesota, moved to New York, and eventually came to the Field Museum in 1983, where I have been a curator ever since. This is my story, as well as that of many natural history museum curators, both past and present.

A captivating and transformational gift from my old friend Hans.

1

Moving toward the Life of a Curator

My thirty-three years as a curator have been filled with exotic adventures, scientific discoveries, and inspiring individuals. A series of chance circumstances and personally influential people set me on track toward a career that has been incredibly rewarding. None of this would have happened if it were not for a captivating gift from an old friend, a compelling university professor in Minnesota, and a few influential curators in London and New York who took an early interest in me.

I grew up in a blue-collar suburb of Minneapolis called Richfield. I lived with my mother, father, four sisters, and a dog—all in a small three-bedroom house with a single bathroom and no basement. I think that is where my competitive spirit first developed, whether competing for time in the bathroom, living space, or prime seating at the dinner table. They were somewhat Spartan living accommodations, but ours was a stable nuclear family with an emotionally supportive environment. My parents had no particular interest in science, but they had no objections to it either. Their support came in the form of encouraging me to follow my heart and embrace the journey. I developed early interests in the natural world because it was an unlimited source of hobbies that I could pursue within my family's limited budget. Through grade school and high school, I collected rocks and minerals from a local gravel pit, raised small freshwater fishes and crawdads from a nearby pond, and collected small fossils from the crushed limestone that paved my parents' driveway. I found great beauty and peace in my personal study of nature, and it kept my adolescent mind occupied.

After graduating high school in 1969, I moved into an apartment to set out on my own. I took on a series of part-time jobs and began taking college classes in search of an occupation that might interest me. My higher education in Minnesota started at Normandale Junior College and later moved to the University of Minnesota. Working my way through college with a series of low-paying jobs in the service industry gave me a respect for the time and resources I spent on higher education. At one time or another, I was a popcorn vendor for a baseball stadium, a busboy for a German restaurant, a cook for an Arby's fast food restaurant, and a weekend drummer/singer for a small-time rock band that played at high school dances and an occasional club. In the early 1970s, I was also a medic in the U.S. Army. Last but not least, there was a year that I spent working in the complaint department of a Montgomery Ward store. Collectively, it was a grab bag of subsistence-level employment to pay for food, rent, and college tuition. The jobs were all character builders of the highest degree and motivators for me to find something better.

By the fall of 1973, I was a junior at the University of Minnesota working toward a business degree. At that point I thought I had figured out what I needed to do with my life; I would either find a job in the retail business like my father or be a teacher like my uncle. Such were the pragmatic goals that had been instilled in me by the working-class environment I had grown up in. The university classes in business were perfectly manageable, but they were not especially engaging to me. I seemed to be plodding along toward an inevitable conclusion.

Then one day in August 1974, something happened that began to change all of that. My friend Hans Radke returned from a vacation in southwestern Wyoming with a souvenir for me. It was a beautifully preserved 52-million-year-old fossil fish in limestone from the Green River Formation. It immediately rekindled my long-harbored interest in natural history. The specimen was so well preserved that it required little imagination to envision it as a living creature, and its unimaginably ancient age further fascinated me. I was enthralled! I couldn't seem to take my mind off of it. So the next day I took my small treasure to a professor of paleontology at the university, Dr. Robert E. Sloan.

Professor Sloan's office was in Pillsbury Hall, a monumental structure made of massive red sandstone blocks that was built in 1887. It was the second oldest building of the entire Twin Cities campus and home to the university's Geology Department. When I arrived at his office, I knocked on the heavily painted panel door and heard a cheerful "Come in." I opened it, peered down a long, dimly lit aisle, and saw him sitting at the far end of the room, sunk into a large padded chair facing me. The office was dusty and cluttered with great piles of books and papers on the floor and tables. I approached him, introduced myself, and handed him my fossil, asking, "Would you please identify this for me?" He took it, stared at it thoughtfully for a moment, and scratched his head. Then he looked at me, smiled, and said, "No. But if you take my course in vertebrate paleontology, you will be able to identify it yourself!" What could I say to that? I signed up a few weeks later.

Professor Sloan was an extraordinary teacher. He was highly animated, waving his arms about when he spoke. He never used notes and clearly loved what he was doing. His specialty was fossil mammals, but he was a storehouse of knowledge about all types of fossils, and he would constantly interject amusing anecdotes into his lectures. I found paleontology, as well as his high-spirited teaching style, to be captivating. I was so taken with it that by the time the course was over, I had decided to change my major from business and economics to geology and zoology. And as promised by Professor Sloan, I was able to identify the fossil fish from my friend Hans. It turned out to be *Knightia eocaena*, an extinct species in the herring family. It is the most common fish species from the Green River Formation and the state fossil of Wyoming.

Once in the geology program, I started learning everything I could about historical geology, paleontology, and the evolution of fishes. I took several more courses from the crafty Professor Sloan, who had hooked me in the first place. I also started visiting the source of my fossil *Knightia* each summer (which I will go into in more detail in chapter 3). Little did I know at the time that this site would later become my most important field site as a professional paleontologist. My interest in fossil fishes grew to include how they fit into the evolutionary network of living fishes and the anatomy of fish skeletons. I received a BS in geology after three years and continued on in a double master's program in geology and zoology. My master's thesis was called "The Paleontology of the Green River Formation, with a Review of the Fish Fauna." During the years of study for my master's degree, I had discovered much about fossils from the Green River Formation, and by 1978 I was making preparations to get my thesis published as a book.

I had learned a lot at the University of Minnesota, but there had been only so much that the professors there could teach me about fishes. I had to delve broadly and deeply into the scientific literature on fossil fishes and fish anatomy for my thesis research and found one scientist doing particularly innovative work in my particular areas of interest: Colin Patterson. Patterson was a principal scientific officer of the Museum of Natural History in London (equivalent to a curator in the United States), and he curated the world's largest fossil fish collection (over 80,000 specimens). He had helped develop an astounding new acid-preparation technique that could make a 150-million-year-old fossil fish appear as though it had died yesterday, providing much more information for scientific analysis. His artistic reconstructive drawings of these fossils seemed to make them come to life (or at least look like fresh dissections of living species). He was also becoming a major authority in evolutionary biology. I decided to seek his advice on publishing my thesis.

I mailed Patterson a draft copy, hoping he would read it and provide some critical review. I wasn't sure that such a world-renowned scientist would agree to do this, but to my surprise, he responded only a few weeks later! Such an immediate response was remarkable in the days when correspondence was written and posted, not sent electronically. He was extremely encouraging, saying that its publication would be a valuable contribution to the paleontological literature. He also strongly recommended that I go to New York and enroll in a PhD program in evolutionary biology under the guidance of Donn E. Rosen and Gareth (Gary) J. Nelson at the American Museum of Natural History. Rosen and Nelson were both curators in the Ichthyology (fish) Division of the museum. Nelson was at the forefront of fundamental change in the world of biological systematics, which is the study of biodiversity, classification, and evolutionary relationships. Patterson was a close, longtime colleague of both Rosen and Nelson, and he convinced them that I might be worth the trouble to take in as a student. A short time later, I received an invitation from Rosen to come to New York with a four-year fellowship covering all costs and living expenses; and so began the most intensive academic training of my life as a student.

The New York graduate program was a collaborative venture between the City University of New York and the American Museum of Natural History, where I was given an office in the museum's Ichthyology Department. My move from the laid-back, working-class suburbs of Minneapolis to the Upper West Side of Manhattan in 1979 was quite a transition. My one-room studio apartment was on the eleventh floor of a high-rise building located at Seventieth and Broadway, only seven city blocks from the museum. The never-ending cacophony of car alarms, horns, engines, and emergency vehicle sirens eventually became white noise in my nights and mornings. I soon came to appreciate the fast-moving culture of the island of over a million and a half residents.

Donn Rosen was the paternal heart of the museum's Ichthyology Department for the staff and graduate students. He loved the American Museum of Natural History, and he was the first person to show me how vested curators can be in training the next generation of scientists. Rosen taught me much about fish anatomy. Fossil species are usually preserved only as skeletons, so knowing the skeletal anatomy of living species enabled me to more accurately identify what I was looking at in the fossils. I also learned from him a critical technique for preparing small alcohol-preserved specimens: clearing and double staining. Two flesh-penetrating dyes color the bones red and the cartilages blue. Then the flesh is rendered completely transparent by soaking the specimen in enzymes and glycerin for many weeks to months. The process results in colored bones and cartilages clearly showing through the body. This technique can be used to produce many convenient-sized skeletons that would easily fit on a microscope stage for careful dissection and examination. While I was in New York working on my doctoral thesis I cleared and stained nearly a thousand fishes and learned to identify most of the major groups of living fishes by their skeletons. The sheer beauty of these specimens further fed my appreciation for the aesthetics and function of anatomy (see page 17).

Rosen had a subversive style of teaching. He taught his students to be unafraid of questioning scientific authority. In a course he gave on systematics, he handed out a list of assumptions that he said *inhibited* progress in comparative biology and evolutionary theory. The list included:

> "Ultimate causes are knowable."
> "Scientists are more objective than other people."
> "Your graduate advisor and/or your distinguished visiting professor are probably right most of the time."

He encouraged students to challenge the system, and he tried to differentiate the more empirical components of science from its more dogmatic beliefs and occasional arm waving. Science, when it works, is an evolving *process* and a testable *method*, not a book of prescribed truths. Rosen lived only to the age of fifty-seven. Nevertheless, he was highly influential in training the next generation of fish systematists, and his PhD students became curators in many of the world's top natural history museums.[1]

A few months after I had moved to New York, Rosen invited me into his office to explain the lay of the academic landscape for my new PhD program. He started by telling me that there was a serious clash of ideologies centered at the American Museum of Natural History involving biological systematics. The controversy sometimes got ugly and spilled over to affect graduate students. He proceeded to explain that after my first year in the program, I would have to pass a four-part written preliminary exam for City University of New York in order to continue on for the PhD. One of those parts was controlled by professors who dogmatically followed a traditional school of systematics with the somewhat pretentious name of "evolutionary taxonomy." This group was adamantly opposed to the school of systematics advocated by Rosen and Nelson called cladistics. The clash of systematic schools was seriously partisan and highly volatile at the time. Rosen explained that I should assume I would either fail or get an extremely low score in the part of the exam that was overseen by the traditionalists because I would be allied with my advisors and other cladists. I would need to take a philosophical stand on the test and defend my point of view, even though there would be a definite cost to doing so. Therefore I would have to give nearly perfect answers for the other three parts of the exam in order to get an overall pass. Rosen's pep talk got my adrenaline pumping.

In preparation for the exam, I spent many long hours reading dozens of books and taking classes on systematics, philosophy of science, evolution, and scientific method. I discussed and debated scientific method with professors, visiting seminar speakers, and fellow students. I also had many hours of in-depth discussions with Nelson and Rosen. They, together with Patterson in London, were some of the most prominent figures in systematic theory and philosophy during my time in New York. Nelson, in particular, was the de facto leader of the cladistic movement.

Cladistics is a search for patterns of relationship among things based on uniquely shared characteristics.[2] Such patterns are called cladograms. When used in an evolutionary context, cladistics is also called phylogenetics. Cladistic method can be used to search for patterns among living organisms, geographic areas, languages, or even non-living objects.[3] When cladograms of species or higher groups of organisms are interpreted as the result of evolutionary process, they are called phylogenetic trees. Traditionally, evolutionary research and paleontology had focused much of its effort on the search for ancestors in the fossil record. It had been assumed by many evolutionary biologists and paleontologists that lineages could be read right out of the rocks, so to speak. Patterson, Nelson, and Rosen all argued that the never-ending search for ancestors had been a waste of time, holding evolutionary research back for decades. An inadequate fossil record together with a number of logistical problems made ancestor recognition from the fossil record impossible. Cladistics, on the other hand, had the power to provide testable patterns of relationship among species. Comparative anatomy (and later comparative molecular data) was the key for scientific analysis of evolutionary relationships. A human and a chimpanzee, for example, share unique characteristics that are not present in a crocodile, indicating a cladistic relationship between the human and the chimp that excludes the crocodile. This does not mean that the human evolved from the chimp, or that the group including the human and the chimp evolved from the crocodile. It is simply a hypothesis of relationship among those species based on uniquely shared characteristics. In an evolutionary context, those uniquely shared characteristics can be explained as having been passed down from a common ancestor, and, logically, the common ancestor of the human and chimp evolved after the common ancestor of the human, chimp, and crocodile (see page 19). Evolution is the only viable natural explanation for the strong cladistic patterns we see in nature.[4]

Unlike more traditional evolutionary trees that specify particular ancestors, cladograms and phylogenetic trees do not have ancestral species at branch points or on the trunk (see page 18). Ancestral species no doubt exist in the collections of museums today, but we simply have no satisfactory scientific method with which to identify them as ancestors to particular species. But we do not need to know ancestors to study evolutionary relationships. I know for a fact that my daughter's direct ancestor is her mother, because I saw her birth. But if you never observed that process in real time, and 50 million years from now you had only the adult fossil of my daughter and the adult fossil of her mother, you would have no way of knowing who was who's ancestor (e.g., which one was the mother and which one was the daughter). Nevertheless, even without knowledge of the exact ancestor-descendant connection, you could still hypothesize a pattern of relationships based on fossils of my daughter, her mother, and a chimpanzee. My daughter and her mother share unique characteristics indicating that the two of them are more closely related to each other than either is to the chimp.

Each new cladogram is a hypothesis of relationship that can be tested, revised, and expanded with additional data and species. The data are empirical in that they consist of observed characteristics from actual specimens. Fossil and living species are treated in exactly the same way in cladistic analyses, with fossil species given no more weight or even less weight than living species when poorly preserved. Fossils simply represent additional species and new character combinations that can add to our understanding of biodiversity and the overall tree of life. The truly remarkable aspect of cladistic analysis is how well characters and species fit together in statistically non-random patterns. The fundamental job of science is to provide natural explanations (theories) for strong patterns in nature. Evolutionary theory[5] provides the only natural (i.e., scientific) explanation for such congruence of species and character data (e.g., there are no reptiles with mammary glands, or plants with vertebrae or inner ear bones). The hierarchical order we find in nature through cladistic analysis is the best scientific evidence we have for evolutionary relationships among species. There is much more to cladistics and phylogenetics than I could possibly cover in this book about curators, but there is a mountain of published literature on the subject.[6]

As a result of learning cladistics, I came to realize two misconceptions I had been taught in Minnesota. The first was the postulate of George Gaylord Simpson and others that "fossils provide the soundest basis for evolutionary classification."[7] The implication was that fossils were more important than living species in the study of evolution. The second postulate was that the goal of evolutionary research should be to identify ancestors. Once I started becoming familiar with cladistics, I realized that both of these postulates were wrong. I began to look at fossils with a more measured eye and appreciate their limitations. Identification of specific ancestral-descendent species lineages based only on a general similarity between species and their relative age in the fossil record is a speculative guess at best. The fossil record is highly incomplete, and only a small fraction of 1 percent of extinct species were ever fossilized.[8] That is not to say that fossils are scientifically unimportant. Extinct fossil species add to our knowledge about the biodiversity that has existed through time, and they further expand the cladogram of life by providing new species and new character combinations. The evolutionary pattern of life on this planet ultimately includes all fossil and living species. But efforts to decipher that pattern benefit no more from fossils species than from living species.

By the end of my first year, I was ready for the preliminary exam. It was a four-hour test, consisting mostly of short essay questions. I finished early and was confident in everything I had written. I couldn't see how anyone could have a problem with any of my answers. A week later, I received the results. Just as Rosen had warned, one of the four sections was downgraded because I had shown a degree of cladistic influence in my responses. No reason was given on the exam—just a big red X over much of my answer. But I had done well enough in the other three areas to more than compensate. I passed. The experience taught me the depths to which scientific controversy can reach in a high-powered academic environment, especially in a time when established tradition is being aggressively challenged.

In the years following the preliminary exam, I increasingly appreciated the pace of intellectual activity at the American Museum of Natural History. It was fast-paced, captivating, and intellectually aggressive. Everyone at the museum was out to prove themselves, and the environment was extremely competitive. You had to be thick-skinned and unafraid to challenge or be challenged. This was cutting-edge science in action. Even distinguished visiting scholars at the museum were drawn into emotionally charged debates. Sometimes a seminar would become a wild forum of shouting, clashing egos, and incompatible ideologies. Scientists in the audience would often speak up or make loud sounds of disapproval in the middle of an invited speaker's presentation. In his 1988 book *Science as a Process*, the philosopher David Hull commented: "Perhaps the seminar rooms of the American Museum are not as perilous as Wallace's Upper Amazon, but they come close." In the same book, he dubbed the aggressive scientific discussions at the American Museum as following the "New York Rules of Conduct." New York rules relied not on status and diplomacy, but on explicit logic and vigorous debate governed by the absence of formality. I soon came to love the pragmatic, irreverently challenging nature of these energetic seminars.

Over the course of my four years in New York, I was immersed in the culture of systematic biology and in the fierce debates within the field. Hundreds of papers were published on the controversies, and scores of scientific meetings hosted loud argumentative discussions about systematic method. Occasionally the arguments became mean-spirited and highly personal. American Museum scientists would often argue their points with one another in published scientific journals rather than simply walking down the hall in the museum to discuss the issues in person. And the controversy extended across the Atlantic. In 1980 Beverly Halstead, a curator of fossil fishes at the British Museum of Natural History in London, published a paper in the international scientific journal *Nature* accusing Colin Patterson of spreading "Marxist" theories, because

Patterson had incorporated cladistic philosophy into a museum exhibit. In other papers, Halstead accused him of "being in bed with the creationists"—strong words indeed to describe Patterson, who was a devoted Darwinian. Some years later, Patterson's nomination for the prestigious Romer-Simpson Medal from the Society of Vertebrate Paleontology was said to have been rejected by influential members of the selection committee because of his views about limiting the significance of fossils in evolutionary studies. Still, Patterson was not deterred. He continued to take a stance against what he saw as a science bogged down in dogmatic ideas. Being out front on this issue made him a target for many groups, including scientists and non-scientists alike.[9]

Gary Nelson was an even more popular target, particularly in the late 1970s and early 1980s. The traditionalist school of evolutionary taxonomy had several highly influential scientists of the day, including two well-known former curators of the American Museum of Natural History, Ernst Mayr and George Gaylord Simpson. Both had left the museum for appointments at Harvard University, which has never been short on clout. Both were prolific writers and iconic figures in their field. (Mayr was still publishing at one hundred years of age.) Both in their later years were major critics of Nelson and of the cladistics method. In 1981 Simpson published a paper in *Nature* warning readers around the world that Gareth Nelson and another one of the museum's paleontologists were among "a small group" at the American Museum "representing a view [cladistics] which is representative neither of the museum as a whole nor the majority of paleontologists." But by this time, the influence of the systematic traditionalists was beginning to waver. In response to Simpson's letter, twenty-two research scientists of the American Museum signed a letter to the contrary that was also published. Twenty-six of the thirty-two biological systematists at the museum were now cladists.

Cladistic philosophy and phylogenetics eventually won over the majority of the scientific community, and the intensity of the "systematics wars" was eventually all but forgotten. Some of my younger curatorial colleagues today are largely unaware of it. Cladistics (known mostly as phylogenetics today) has largely replaced the so-called school of evolutionary taxonomy in systematic research. Experiencing the heart of the controversy and fundamental shift in philosophy gave me an appreciation for the process of science that I will never forget. The media is good at publicizing the significant discoveries of science, but often not as good at documenting the long process of paradigm shifts that lead to those discoveries. It takes living through such transitions to fully appreciate their importance.

The American Museum of Natural History was the best place in the world for me to be for my doctoral education. I learned an enormous amount about scientific analysis, specimen preparation, and fish anatomy. Perhaps most importantly, I learned firsthand about the dynamic struggles that are sometimes required for science to move forward. By the end of my third year there, I was sure that I wanted a career as a research curator in a major natural history museum.

Colin Patterson, principal scientific officer of paleontology for the Natural History Museum in London, on a nature walk through London in 1985. His recommendation to Donn Rosen at the American Museum in New York put me into a life-changing PhD program.

The acid-transfer fossil preparation I learned from Colin Patterson was a breakthrough for paleontological research. (*Top*) 150-million-year-old fossil bowfin fish, *Solnhofenamia elongata*, in original limestone slab with crude mechanical preparation. (*Middle*) The same specimen after I embedded it in epoxy resin and gently dissolved the hard limestone from the other side, producing a pristine surface of 150-million-year-old bones in the clear hard resin. (*Bottom*) A cleared and double-stained skeleton of the living bowfin, *Amia calva*, for comparison to the fossil.

Curator of ichthyology at the American Museum of Natural History Donn Rosen (*right*) and me at the 1982 meeting of the American Society of Ichthyologists and Herpetologists. Donn was the paternal heart of the museum's Ichthyology Department for both staff and graduate students. For more on his personal history, see pages 364-65.

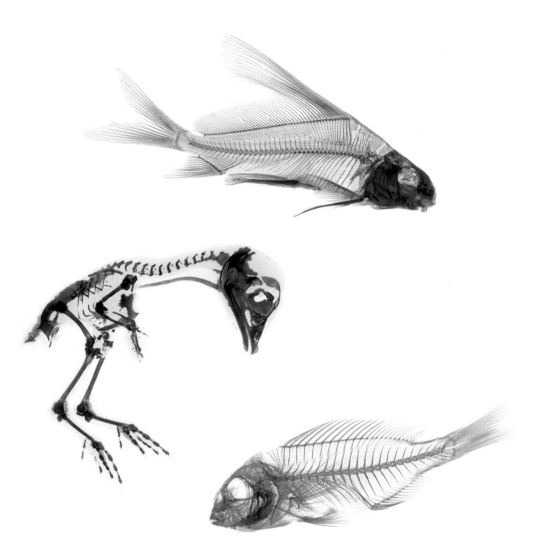

Beautifully cleared and double-stained skeletons of a chick and two small fishes. The aesthetics of anatomical form is part of what drew me to comparative anatomy. My graduate training with Donn Rosen taught me the clearing and staining technique among other things that helped me become a better anatomist.

From traditional evolutionary trees to cladograms: A move toward scientific empiricism.

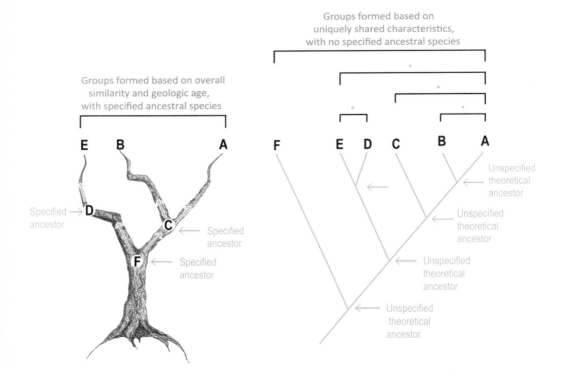

Evolutionary relationships of A-F represented as a traditional tree

Evolutionary relationships of A-F represented as a cladogram

(*Left*) Traditional tree scenario with its designated ancestral species (**F**, **C**, and **D**). Specific ancestral species were once identified using a combination of overall similarity and the relative geologic age of fossils. (*Right*) With the advent of cladistics, evolutionary studies became more empirical. The goal became to determine a hypothetical pattern of relationships based on special shared characteristics (e.g., species **A** is more closely related to species **B** than to species **C**) rather than identifying specific ancestor-descendant species lineages (also see cladogram on page 19). There is currently no rigorous scientific method to identify specific ancestral species, especially given the incompleteness of the fossil record.

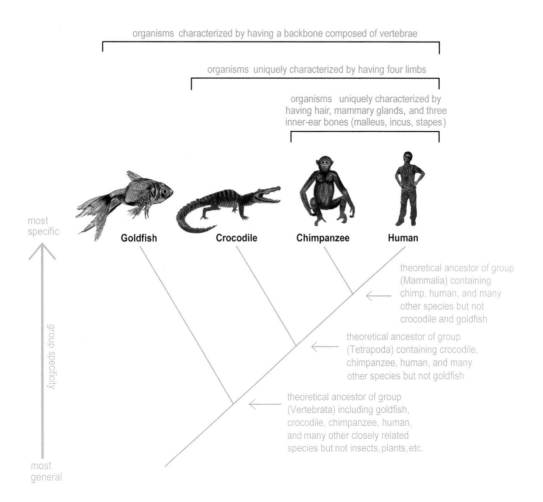

Cladistic pattern of relationship (*blue lines and text*) among four select organisms based on shared, uniquely derived characters (*red text*). Even if we only have four distantly related species (a human, a chimp, a crocodile, and a goldfish), we can still arrange them in hierarchical groups of relative relationship based on uniquely shared characteristics. As more groups (e.g., species) and characters are added to this branching diagram, it becomes increasingly informative as a network of relationships among species. Evolutionary theory provides the only natural (i.e., scientific) explanation for such high congruence of species groupings and character data. The job of science is to focus on natural explanations for repeating patterns in nature.

Gary Nelson, curator of ichthyology at the American Museum of Natural History (*right*), and the scientist whose work inspired him to lead the cladistic movement, Lars Brundin, in Stockholm in 1988. They were two of the most influential systematic biologists of the twentieth century. For more of Gary's personal history, see pages 363-64.

New York's American Museum of Natural History, founded in 1869, is located on park-like grounds in the heart of the Manhattan's Upper West Side. Today the museum is a complex of twenty-seven interconnected buildings with a total floor space of about 1.6 million square feet. My four years as a doctoral student there taught me much about the life of a museum curator.

Chicago's Field Museum of Natural History, on the southern shore of Lake Michigan, where I began my curatorial career. The museum was founded in 1893 (opening to the public in 1894) at another location six miles south of here and moved to this location in 1921.

2

Beginning a Curatorial Career

In the spring of 1982, while I was at a scientific conference in DeKalb, Illinois, David Bardack alerted me to a curatorial job opening at the Field Museum in Chicago. Dave was a professor in the Biology Department at the University of Illinois at Chicago and a research associate of the museum, and we knew each other through shared research interests about fossil fishes. (The fossil fish world is a close-knit one.) He said that the Field Museum had been searching for a curator of fossil fishes for years, and although they had interviewed several applicants, they had been unsuccessful in finding an acceptable candidate. The current job advertisement specified an invertebrate paleontologist, but Dave thought if the right sort of fish paleontologist applied, he or she would have a good shot at the position. Even though I was still a year away from finishing my doctoral thesis in New York, I figured that I had nothing to lose by applying. In fact, this was a rare opportunity. How many curatorial jobs in paleoichthyology (fossil fishes) could there possibly be after all? I sent in my application.

A few weeks later I received a letter from John Bolt, chair of the Geology Department and curator of fossil amphibians and reptiles at the Field Museum, inviting me to come to Chicago for an interview. Rosen and Nelson were both happy that another of their PhD students might possibly find employment. My fellow museum graduate students were delighted that the competitive pool of job-seeking ichthyology graduates might decrease by one. And I was ecstatic about the possibility of landing a career-track research job without having to go through a series of one- to two-year-term postdoctoral positions around the country. I had not been looking forward to the rootless postdoc stage of an academic career that has often been referred to as the life of a gypsy scholar. I arrived in Chicago in July 1982 ready to give two public presentations and to go through two days of one-on-one interviews for the position of assistant curator. This was to be my first experience interviewing for a professional scientific job, so on the flight to Chicago I was a bit nervous. But once the plane landed at O'Hare, something about coming back to the Midwest instinctively put me at ease.

On the morning of my first day of interviews, I took a cab from my hotel and arrived at the museum around 9 a.m. The beauty of the Field Museum on the sunny shoreline of Lake Michigan was exhilarating. There was even a gentle breeze off the lake that day that put a remarkably fresh smell in the air for such a densely populated urban location. It was quite a change from midtown Manhattan. I met John Bolt at his office. He immediately created a friendly atmosphere with his good-natured humor and warmth. Bolt was a third-generation midwesterner who had done his doctoral work at the University of Chicago, just a few miles down the road from the museum. After graduating, he had spent four years as an assistant professor of anatomy at the University of Illinois before being hired as assistant curator of fossil amphibians and reptiles at the Field Museum in 1972. It became clear through his introductory interview with me that he loved curatorial work at the museum and that he had a great vision for the future of paleontology in the Geology Department.

At 10 a.m. I gave a presentation to the four scientific departments of the museum (geology, botany, anthropology and zoology) about my research. It was a forty-five-minute presentation followed by fifteen minutes of Q&A. It seemed to go well, although as it proceeded I became worried because the audience seemed so quiet and restrained. My years of giving and attending presentations at the American Museum of Natural History had accustomed me to more overtly challenging spectators. In New York, if at least one person wasn't challenging you during your talk or notably upset about something you said in your conclusion, you usually figured that people weren't really listening to you. And you could often tell how people felt about your words of wisdom by the strained expressions on their faces. But Chicago was different. The faces in the audience were more restrained and harder to read, and the two people that had the most challenging questions during my talk waited until after the question-and-answer period of the public presentation to quietly come up to ask their questions in private. This level of civility would take some getting used to.

For the rest of the day, I had a series of interviews with each of the curators. The collection manager showed me the fossil fish collection. It was an impressive one, with more than 15,000 specimens at the time. This would nicely complement the collection of 1.5 million alcohol-preserved fishes in the zoological collections on the ground floor. I then met the staff of fossil preparators. The Field Museum preparators are true artists of the trade. The quality of their work on exposing fossil bones encased in rock was amazing. Using needle-sharp tools under a microscope for thirty-five hours per week or more, they managed to expose even the tiniest, most delicate bones without damage. I was already imagining how I could take advantage of this amazing pool of talent if I got the position, together with the acid-preparation techniques I had learned in London and New York. It was a dream job for anyone contemplating a career in specimen-based research on fossils.

Another attraction of the Field Museum was that, like the American Museum, the Field Museum Research Division was an academic institution at its core. The Field Museum had formal collaborative programs with three local universities: the University of Chicago, the University of Illinois at Chicago, and Northwestern University. The collaborative arrangement included hosting dozens of graduate students in offices within the museum, curators teaching university courses at the museum or on the university campus, and curators acting as mentors and graduate advisors for students. The main collaborative PhD program whose faculty I would be a part of was with the University of Chicago's Committee on Evolutionary Biology (CEB). The Field Museum was one of the founding partners of this program in 1968, and it had developed a reputation of international acclaim. Curators are academics at heart, with the exact same level of educational training as college professors. And as I learned in New York and later confirmed in Chicago, resident graduate students are the lifeblood of any active research institution. They help keep new ideas flowing into the scientific research programs, provide fresh energy for moving projects forward, and keep museum curators firmly connected to the academic community.

Another major benefit of a curatorial position at the Field Museum was the possibility of tenure, or what is sometimes referred to as "career status." Curatorial positions at the Field Museum were like curatorships at the American Museum of Natural History and professorships at all major universities in that they were part of a tenure-track system.[1] Tenure serves as a protection of academic freedom and an encouragement of innovation. Tenure is also a perk to attract the best scientists to an institution. When it works properly and is used strategically, modern tenure is an effective method of putting newly hired researchers to the test before making a career commitment to them. It amounts to a strict five-year probationary period. If by the sixth year the assistant curator of a major research museum (or assistant professor of a typical university) has not demonstrated a measurably high level of performance and passed a long and thorough review process, he or she is fired. There is no middle ground. During the four years prior to my interview in Chicago two assistant curators in the Geology Department had failed to pass the tenure review process and had been let go. But if tenure is achieved, it allows a scientist to engage in long-term projects that may take years to come to fruition but have a commensurate impact in the end. It also empowers younger scientists to challenge the older established generation of scientists who, in some cases, may have become dogmatic and reluctant to change. The overall effect of tenure is to help science progress.

In 1982 I was not yet sweating the challenge of the tenure review process. My priority then was to get hired and have a chance to prove myself. At the end of the second day of interviews, I was tired but energized by the feeling that I had done well. I returned to New York the next morning. The hiring committee members also seemed impressed with the fact that I already had published a successful book going into a second edition (the publication based on my master's thesis) and that I had developed a highly productive field site in Wyoming.

A few days later I received a phone call from John Bolt offering me the job at a modest salary contingent upon my finishing my PhD. I thought about it for half a millisecond, and then I said, "Sure!" Bolt kind of chuckled and then asked, "Well, don't you want to negotiate? I mean I could probably get you an extra $500 per year plus some additional start-up funds." I responded, "Oh yeah, that sounds really great!" I still had a lot to learn about the fine skills of negotiating (skills I have since refined). Nevertheless, all this was good news. As far as I was concerned, it was the perfect job in the perfect location.

By the time I started working at the Field Museum in late September 1983, the curatorial staff of the Geology Department at the Field Museum was in the early stages of a major transition period thanks to John Bolt.[2] He was building a world-class Paleontology Department and determined to hire scientists who would become leaders in their respective disciplines. In 1981 the department had been more of a traditional, university-type geology department, including curators who specialized in mineralogy, petrology, and sedimentology.

But under Bolt's leadership, the department refocused on its greatest collection strength: the fossil collection. Bolt also broadened the geographic scope of the departmental hiring searches. In 1981 half the curators in the department (including Bolt himself) were PhD graduates from the University of Chicago. Although the University of Chicago is a top graduate school, any faculty made up of too many graduates from the same institution risks becoming functionally inbred and limited in originality. Under Bolt's leadership, the curatorial hires over the next eight years came from broader international searches, with degrees from all over the world. In 1982 Peter Crane[3] (PhD University of Reading, United Kingdom) was hired as curator of fossil plants. In 1983 I (PhD from City University of New York) was hired as curator of fossil fishes. In 1984 Scott Lidgard[4] (PhD Johns Hopkins University, Maryland) was hired as curator of fossil invertebrates. In 1988 John Flynn[5] (PhD Columbia University, New York) was hired as curator of fossil mammals. And in 1990 Olivier Rieppel[6] (PhD University of Basel, Switzerland) was hired as curator of fossil reptiles and amphibians. The primary expectation of us when we were hired was clear: to be productive scientists. We were charged to make new scientific discoveries, write successful grant applications, produce significant publications, and train doctoral or postdoctoral students to help build the next generation of evolutionary biologists. We became professionally established while taking over from the preceding generation of departmental curators. The five of us formed an especially synergistic group in what was a golden age for curatorial science in the United States and in the Field Museum as well.[7] The bonds of friendship and brotherhood among us are the sort that can only develop over time and from shared history, and they remain intact, even as some of us have gone our separate ways to other institutions.

 The Chicago region became my permanent home, and my curatorship at the Field Museum began a long and rewarding career for me. Some combination of my love of the Midwest, the collegial atmosphere of the museum, and the thrill of a new opportunity immediately hooked me at the deepest level. Like the other curators in Chicago, I had a passion for collection-based research, writing, and teaching in an institution that fostered those activities. And I had one other thing that helped me get the job in Chicago in the first place: a well-established field program in one of the world's most productive fossil localities. That had turned out to be a strong asset for me during my interviews for the job, and it has been a constant asset throughout my career. Fieldwork is one of the most important ways that curators make scientific discoveries and add to museum collections—which brings me to another chapter in the development of my career as a scientist: staking out a field site. That process began for me long before I got the job in Chicago, and even before I went to New York. It began while I was still a geology student at the University of Minnesota.

The curators of the Field Museum's geology department in 1994. *From left to right:* Lance Grande, John Flynn, Olivier Rieppel, Peter Crane, John Bolt, Scott Lidgard, and Matt Nitecki. Image taken on May 4, 1994, in the *Life Over Time* exhibit of the Field Museum.

(*Top*) Peter Crane, curator of paleobotany at the Field Museum from 1982–99, standing between me and Speaker of the House Newt Gingrich in 1995. (*Bottom left*) Peter at Kew Gardens giving a personal tour to the queen of Great Britain, Elizabeth II, in 2004. (*Bottom right*) Peter with his wife, Elinor, in Tokyo, Japan, where Peter accepted the International Prize for Biology from the emperor and empress of Japan in 2014.

Olivier Rieppel, the Rowe Family curator of evolutionary biology (*left*), with paleontologist Eitan Tchernov of the Hebrew University holding the acid-transfer prepared specimen of *Haasiophis terrasanctus*, a 93-million-year-old fossil snake with legs, from the Middle East. Olivier has worked on several such transitional forms (species with transitional character combinations) over his career.

(*Top left*) Scott Lidgard, curator of invertebrate paleontology. (*Top right*) John Flynn (*at left*), curator of fossil mammals at the Field Museum from 1988 to 2004, on an expedition in the mountains of Río Las Leñas, Chile. (*Bottom left*) John Bolt, curator of amphibians and reptiles, with a 340-million-year-old skull of *Watcheeria deltae* that he found in Delta, Iowa. It is one of the earliest-known land vertebrates. (*Bottom right*) Matthew Nitecki, curator of invertebrate paleontology from 1965–96.

The high mountain desert of Wyoming, where I found my first and most successful field site, and where I have worked now for four decades.

3

Staking Out a Field Site in Wyoming

In early September 1975, only a year after I had transferred into the geology program at the University of Minnesota, I set out on my first trip to find the source of the fossil fish my friend Hans had given me. The small *Knightia* specimen that had changed the course of my career aspirations came from the high mountain desert region of southwestern Wyoming, about 12 miles west of the small town of Kemmerer. There, at around 7,000 feet above sea level, is a limestone layer in what is called the Fossil Butte Member (here abbreviated as FBM) of the Green River Formation. Encapsulated within it is an entire 52-million-year-old community of extinct plant and animal species, as though frozen in time. Billions of beautifully preserved fossils there reveal an unusually graphic window into Earth's distant past. I still clearly remember the anticipation and excitement I felt on that first trip.

I went with my friend Hans and his brother-in-law Rick Jackson. They had both been there the previous year and were happy to show me what they had seen and experienced. We slept in tents on the outskirts of Kemmerer in a campground/trailer park on the Hams Fork River. The nights this time of year in the high mountain desert were cold and below freezing, even though the daytime temperatures were usually in the 70s or 80s. That first night sleeping was a challenge because of the adjustment to breathing oxygen-deprived air at 7,000 feet above sea level and the unexpected cold (it went down to 17°F!). Sleeping was further disrupted because our tent site was next to a railroad yard where the sound of idling diesel engines could be heard at all hours of the night. The ground periodically shook like a small earthquake when large moving coal trains rolled by. But at least the campground had drinkable water, a restroom, and even washing machines for laundry. For the cost of an out-of-state fishing license, you could catch fresh rainbow trout out of the Hams Fork River for dinner. And you were only minutes away from town for lunch, supplies, and killing time if it was raining too hard to drive up the steep muddy road to the fossil quarry.

On the first morning in Kemmerer after our frigid night in the campground, Hans, Rick, and I went to visit one of the most famous commercial fossil quarries of the time, the Ulrich quarry. Fossils from the FBM have been commercially mined for over 150 years. The vast majority of fossils excavated from these sites have come from commercial collectors. When well prepared, the beauty of these fossils is such that they are in demand by serious fossil collectors and as art objects by natural art dealers and interior decorators. There are even companies that make high-end countertops and fireplace mantles out of FBM slabs of stone containing the common species of fishes. By 2014 there were at least ten different commercial enterprises mining fossils in the FBM. The deposits are extensive and cover an area of hundreds of square miles. This means that this fossil locality is effectively unlimited because a single 8-by-10-foot section only 18 inches thick can take a crew of twenty a month or more to properly excavate. There are enough rich fossil deposits to last for centuries, perhaps millennia, even with the steady commercial exploitation.

Carl Ulrich, his wife, Shirley, and their son Wally were part of a long line of commercial fossil quarriers in the FBM dating back to the late nineteenth century. Carl had offered to show us the techniques he used to excavate fossils and to let us dig in his quarry. We would not be allowed to keep much of what we found unless we were willing to pay for it, but I saw the whole exercise as a learning experience. I could observe how the Ulrich family excavated fossils and learn from their experience of the last several decades. Carl had a commercial permit for digging fossils on state land. The permit cost next to nothing at the time, and the terms of the permit were that he could keep the six most common genera of fossil fishes, fossil plants, and fossil insects, but he was required to turn over all the other fossils (the so-called "rare vertebrate species") either to the University of Wyoming or the Wyoming State Geological Survey based in Laramie. In theory it was a policy in which everyone (the quarrier and the state) would benefit, although it was based strictly on the honor system. We worked for several days in the Ulrich quarry and found several hundred fossils. We were given a few of the broken pieces and a couple nice specimens of the most common species, *Knightia eocaena*, to keep. The other fossils we found were available to us for purchase if we wanted to buy them. The excitement of lifting slabs of rock and seeing animals and plants that had not seen the light of day for 52 million years fueled my growing interest in paleontology. Digging fossils in one of the world's most productive fossil sites is like nothing else in the world.

After that first trip, I continued to go back to Wyoming each year to do more fieldwork. The more that I learned about the FBM fossil locality, the more it became apparent how little was known about it and how much the locality had to offer. I saw it as more than just another fossil locality. I saw it as an entire ecosystem whose biodiversity could provide a uniquely complex documentation of Earth's deep past. It was still largely unexplored, and I needed to make a long-term commitment to decipher the historical information that the FBM had to offer. A major challenge for me during those early years was that I had to pay for most of the pieces that I collected in the Ulrich quarry. I spent hundreds of dollars each year purchasing common fossils that I had found myself, and I could not afford to acquire any of the scarcer varieties. Once I brought the pieces back to Minneapolis, I prepared them myself with needles under a microscope. In one of Sloan's vertebrate paleontology classes during my second year in the geology program, I gave a presentation using the Wyoming fossils that I had found, bought, and prepared. I started to realize that specimen-based research was something that I would enjoy doing for a career. But by the third year of going to Wyoming, I could see that as a poor college student I could not afford to continue paying for the fossils that I found in the quarry. I needed another way to continue building a research collection. Then I came up with an idea.

I had learned from working in the Ulrich quarry that the 18-inch-thick prime fossil layer was almost exactly horizontal and occurred at about 7,000 feet above sea level. The 52 million years of normal Earth deformation had not appreciably tilted the FBM deposits. So I searched a topographic map for nearby elevations of 7,000 feet, reasoning that the fossil layer should be present at those locations. I then transferred the coordinates of the potential fossil sites onto a land-use map, which shows landownership. I discovered many potential new sites, as well as the names of individuals or agencies I needed to talk to in order to work those deposits. As a student, I did not have the desire to open a commercial quarry nor did I have the resources to hire bulldozers and equipment. Luckily for me, Rick Jackson did. He had been dreaming of having his own quarry for years but lacked the know-how to find a productive site. So I struck a deal with him. I would show him locations that I knew would be productive, and he would file for the state permit, pay all permit fees, and pay for the bulldozers to remove the 20 feet of overlying rock. I would have the right to dig in his quarry as long as he had it, and I could keep half of whatever I found. He would also throw in any pieces that appeared to have special scientific value. It seemed like a great deal to me, and I would no longer have to pay for fossils I found, including scarcer species that I had not been able to afford to purchase from the Ulrichs.

We opened the quarry on state land in 1978. We moved our camp from the campground in Kemmerer to the butte top just above the quarry. As I expected, the quarry turned out to be an extremely productive site. Rick was true to his word and honored his agreement with me each year without fail. I began accumulating a diversity of fossils, including fishes, insects, snails, and plants. Rick was a good-natured soul who loved fossils, family, and cheap midwestern beer. He had a fossil preparation lab set up in his house, much to his wife's dismay. Fossil preparation is a dusty business. In Wyoming, Rick loved to stargaze at night and sit around the campfire drinking coffee and smoking cigarettes. He was a hard-core caffeine addict. He drank four pots of coffee a day, plus a six-pack or two of Coke. He claimed that he could not fall asleep at night without first drinking his evening pot of coffee in its entirety. I am not sure what was worse for him; all of that caffeine, the two packs of cigarettes he smoked each day, or all of the dust he inhaled over the years preparing fossils without proper ventilation. Some or all of these habits would eventually contribute to his early death in 2001 at the age of fifty-eight. I am glad to be able to say something about Rick here, because he played a role in my career development and I wouldn't want him to be forgotten from any account of my start in the FBM.

Rick's operation in the FBM only lasted a few years. By the early 1980s, he had begun to face a challenging cultural environment. The competition between the commercial quarries in Fossil Basin back then could be brutal. Some of the older quarriers resented out-of-state upstarts like Rick who had undercut their prices on fossils and drove previously established market values down. They began to see him as a competitive threat to their livelihood. There were fictitious reports sent to Wyoming State officials by anonymous informants about Rick abusing his permit site, and once there was even a threat against his life by a competitor. After a bad experience with poachers who vandalized and stole fossils from the quarry, we had to leave guards in the quarry when we went to town to get supplies. Rick commuted to his Wyoming quarry each summer from his home in Minnesota. The trouble was, he could only be there for a few weeks each year. He couldn't keep up with the quarry vandalism that occurred while he was back home in Minnesota. This left a mess that the public officials finally found unacceptable, and in 1982 the state of Wyoming revoked his permit.

By 1983 I had moved to my first curatorial job in Chicago with visions of building the world's best FBM collection there. The loss of Rick Jackson's quarry permit was a setback for me, but I was undaunted. My interest in the FBM had only increased over time. I could have acquired a scientific permit from the state of Wyoming at any time, but in the 1980s Wyoming was much more generous to commercial quarriers than to scientific ones. The Wyoming commercial permit, which cost next to nothing at the time, gave title for the fossils to the person who quarried them, while the Wyoming scientific permit would only give the fossils on long-term loan to the scientist's institution from the state of Wyoming. Partnering with a commercial quarrier was my best option, but after my experience with Rick, I realized that I needed to partner with a Wyoming resident who would oversee a quarry for the entire season.

I still had my map of potential FBM localities, and some of them showed up on private property that I reasoned might be less problematic than setting up another state quarry. A private ranch owned by the Lewis family of Fossil, Wyoming, appeared to be the best prospect, based on my maps and preliminary explorations. So I built a new cooperative group that included me, the landowners, and a third-generation commercial fossil quarrier from Wyoming named Jim E. Tynsky. I would show Tynsky the most promising place to prospect for a quarry site, and I would be a consultant on its initial development. Tynsky would pay all leasing fees to the Lewis family for a five-acre quarry site, provide all bulldozing, and keep the quarry occupied all year. It was a gamble for everyone involved, because the initial investment was steep as were the leasing fees, and there was no guarantee that the highly fossiliferous 18-inch layer would be minable there. But if it worked, I would have digging privileges for life—or at least as long as Tynsky held the lease—and I would get half of what my crew and I found each year for the museum. I could select pieces that would be most useful for the research collection, and the museum would have clear title to them.

In the summer of 1983, we bulldozed down through 20 feet of overlying rock to a zone where I predicted we would uncover the fossil-rich 18-inch layer. We hit the paleontological jackpot. By 1984 it was clear that we had found one of the richest fossil sites within the FBM, and this is the site I still work at to this day. For two or three weeks each summer over the last thirty-one years, I have brought a field crew of twelve to twenty Field Museum staff, students, and volunteers to work the site. There we have collected thousands of beautifully preserved fossils for the museum collections, discovered dozens of new species, and assembled a uniquely graphic look at Earth's deep past. Although I could have had a scientific permit on state or federal land at any time over the last several decades, the museum could still not have obtained clear title to specimens excavated there through those permits. Since coming to the Field Museum, I have had to oversee a collection which has grown to over 20,000 fossil fishes, and I have not wanted the extra complication of cataloging specimens on loan from another institution or state office into that collection. Partnership with a private quarry has continued to be the most practical way to go.

Developing a field site is like many other important things in life; you build it over time, layer upon layer. In the very beginning, my goal was preliminary exploration of the FBM localities, but over the years it has become to better understand the complex biodiversity of the 52-million-year-old community. I quickly realized that in order to obtain an adequate sample size for my research, I needed to build a complex network of personal connections with people, places, and organizations. So gradually I built a programmatic consortium including FBM landowners, stone quarriers, fossil collectors, commercial fossil dealers, local universities, museum curators, National Park Service employees, and public officials from the state of Wyoming. Eventually, I had developed a large network of commercial and amateur fossil collectors who would notify me when something new or unusual was excavated in one of the FBM quarries. I became a great advocate of a citizen-science approach to paleontology. This approach is a crowd-sourced method of resource collecting that harnesses the efforts of large numbers of amateurs and non-scientists. For me it has become like having hundreds of full-time field-workers mining the quarries for six months every year. It is also a great way to engage the general public in the scientific process, which helps them better appreciate its importance. This approach would not work in every paleontological locality (e.g., localities with very limited deposits), but the geographic extent and richness of the FBM is so vast that there was no other way to effectively explore it. I also engaged a group of loyal volunteers in Chicago who helped with fossil preparation and assisted with cataloging of specimens. Today there are so many interlinked parts to my operation in the FBM that it sometimes amazes me.

Over the years, I have met most of the people who have worked in the ten or more active commercial fossil quarries within the FBM. These quarry operations mine fossils from the time the snow melts in late May to the first heavy snows of October every year. Each quarry has from two to more than a dozen workers and an unknown number of volunteers who dig for shares, totaling hundreds of people mining fossils in the Fossil Butte Member for five or more months every year. This is where the citizen-science concept became particularly useful in my study of the FBM's past biodiversity. My crew of anywhere from twelve to twenty people working for two weeks per year (with two days off) totals only between 120 and 200 workdays per year digging for fossils. The commercial quarries total over 20,000 workdays per year. Many of these quarries are constantly on the lookout for unusual specimens that I ask them to watch for, and they notify me when they find these fossils. This large network allows me an unprecedented look at the diverse assemblage within the FBM. In a locality like this, a large sample size of fossils is critical because of the scarcity of many species. For example, specimens belonging to the pike fish group (Esociformes) occur at a rate of less than one out of every 1 million fish specimens. If my museum crew and I worked every summer for the next ten centuries, we could not expect to recover a million fishes from the FBM. More than 99 percent of the FBM fossils that have been excavated over the last 150 years have come from commercial and amateur fossil quarriers.

Commercial versus scientific collecting of fossils from important paleontological sites is a sensitive issue. The most extreme opinions range from those who think commercial excavation of fossils should be openly permitted on all lands, public and private, to those who think commercialization of fossils should be prohibited because it represents a threat to science. The reality of the situation is that each fossil locality needs to be considered on a case-by-case basis. Some sites should be protected, such as small limited deposits of important fossils on federal lands. Others, like the FBM, could not possibly be adequately sampled for scientific study without the help of amateur and responsible commercial interests. Museums and universities will never have the resources to adequately explore such sites alone. There could one day even be a *need* for the help of amateur and commercial excavation for salvage reasons. Mining interests once proposed strip-mining oil-rich rock underneath some of the fossiliferous deposits of the FBM to produce oil. To do so profitably would have required the destructive removal of overlying rock containing the 18-inch-layer FBM.

Commercial fossil collectors and professional paleontologists have not always been at such odds with one another. The earliest commercial quarriers of FBM fossils date back to the 1870s. They are known today by such colorful names as "Pap Wheeler," "Stovepipe Smith," and "Peg-leg Craig." These hardy souls led a hard life, quarrying fossils for a living without the use power saws, tractors, or trucks. Some, like Peg-leg Craig (who only had one leg), struggled to push wheelbarrows of tools and fossils up and down the steep butte face, and some ultimately died of injuries sustained in the field. They were pioneers of sorts, who sold their fossils to museums in the East such as the Yale Peabody Museum, the American Museum of Natural History, and the Smithsonian.[1] While it is true that some commercially excavated specimens of unique importance sometimes end up in private hands, the majority of these end up in museums. The net effect is that many more unique specimens have been obtained by museums with the help of commercial collecting than would have been possible without it. Another frequent complaint about commercial quarries is that some do not record adequate locality data for the pieces they collect. That is not so much a concern for the FBM material because it comes from relatively narrow, easy-to-identify stratigraphic horizons (i.e., the rock that contains the fossils is diagnostic of the locality from which it came). I can almost always tell where a specimen was originally excavated by examining the rock and/or knowing the quarrier who found it. The cooperation I have received over the years from the commercial quarriers and amateur collectors in the FBM has aided my work immensely. The variety of fossils I have been able to accumulate and study has provided an unusually comprehensive picture of a 52-million-year-old ecosystem. And in spite of the millions of fossils that have already been recovered over the last century and a half, I still see new species of FBM plant or animal fossils discovered every year.

The FBM assemblage of fossils illustrates an important chapter in Earth history. It is the best documentation we have of North America's early recovery after the great Cretaceous extinction event of 65 million years ago. That cataclysmic event, caused by a massive asteroid impact on the coast of the Yucatán Peninsula, wiped out as many as three-fourths of all living species on Earth. It caused the total extinction of dinosaurs (with the exception of birds), marine reptiles (mosasaurs and plesiosaurs), flying reptiles (pterosaurs), ammonites, and many other groups of animals and plants. But Earth continued to evolve. The mass extinction cleared the way for a great evolutionary diversification of flowering plants, pollinating insects, mammals, birds, and other groups. The FBM is a graphic documentation of how the world's biosphere was reshaping itself some 13 million years after the asteroid took its toll. All of the FBM species have been extinct for millions of years, but many of them are early representatives of families and orders that diversified over time and flourish today.

The preservation of the FBM fossils is exceptional, with complete skeletons including scales, feathers, skin impressions, and occasionally even color patterns. Hundreds of extinct species of plants and animals, from microscopic pollen and insects to 12-foot palm fronds and crocodiles are preserved together as an entire 52-million-year-old subtropical community locked in stone. The FBM contains much more than isolated examples of extinct species. There are entire growth series for most of the fish species, from specimens still coiled up in eggs to large adults. Some species even show stages of reproduction and birth, most notably in the stingray *Asterotrygon*. One slab of this rare species (only about forty specimens were known as of 2013) has a male and a female specimen preserved in copulatory position (yes, sex in the fossil record!). Another specimen shows a pregnant female with an embryo visible inside of her, while another slab shows a female that was preserved just moments after giving birth with two babies fossilized beside her. I suspect that this species normally lived in connecting tributaries, and they swam into Fossil Lake for the purpose of reproduction. We can identify other behavioral and ecological aspects of the Fossil Lake community as well. Schooling behavior in fish species is clearly documented by mortality slabs with the entire schools preserved en masse. Well-preserved stomach and mouth contents and fossilized fecal material reveal the diets of many species. Leaves with insect chew marks along with the insects that made the marks are preserved. There is even a fossilized record of diseases and injuries among the FBM organisms. The detailed record of anatomy together with that of growth, behavior, and pathology serves to bring this assemblage alive, and we can study it as we would study an interactive, living community. The graphic diversity of fossils is illustrated and discussed in detail in another book of mine, *The Lost World of Fossil Lake*, published in 2013.

A large portion of the FBM was established as a National Monument in 1972, establishing the site as a piece of national patrimony. In the late 1980s, the National Park Service asked me to help develop a new museum to be built as part of the monument. The original building for the monument set up in the early 1970s had been a converted single-wide house trailer with minimal furnishings and a leaky roof. The Park Service had come to regard the FBM to be important enough to warrant a significant upgrade for its visitor center, and Congress agreed, approving a multimillion-dollar proposal for that purpose. For the next two years, I worked in my off hours from the museum as a contractor with the Park Service's Interpretive Design Center out of Harpers Ferry, West Virginia, and with sedimentologist Paul Buchheim from Loma Linda University in California to help create a flagship paleontological museum for the monument. This included designing a layout for exhibit cases, working with a mural artist for reconstructed FBM scenes, purchasing fossils from the local quarries for exhibit, making casts of some large Field Museum specimens for the exhibit, and producing some interpretive videos to be shown daily at the monument. The project took two years, and in 1990 the new building opened. It was a beautifully designed, eco-friendly building set in a valley surrounded by buttes containing the FBM, and it has been a major attraction of southwestern Wyoming ever since. This brought one more entity into the network of my FBM program: the growing paleontological staff of the National Park Service (NPS). Over the years, the NPS has sent many park interns and staff up to work with my field crews in the Lewis Ranch quarry. They also work with the local quarriers to get some of the important new finds each year incorporated into the collection at the monument or at the Field Museum. The NPS paleontologist Arvid Aase has done a remarkable job in cultivating connections between the commercial quarries, the Field Museum, and the National Park Service.

Throughout my curatorial career in Chicago, I have had many different research projects and publications focusing on a variety of biological, geological, and even philosophical topics; but many of those projects eventually connected to some aspect of the Fossil Butte Member fossils, where I first made my name in paleontology. The last thirty-three years of my fieldwork in the FBM have expanded the Field Museum's collection of FBM fossils to be the largest, most diverse of any museum by far. It includes thousands of specimens that will be preserved in perpetuity for study by scientists around the world. In 2006 I used over a hundred of these pieces to help design a permanent gallery in the Field Museum's *Evolving Planet* exhibit, which is seen by over a million visitors annually. In 2013 my book *The Lost World of Fossil Lake* received the PROSE award for the best book in Earth Sciences. The FBM has been good to me.

For the last thirteen years, I have also taught a field paleontology course there called "Stones and Bones" through the University of Chicago Graham School, combining paleontological excavation for my research with educational training of highly motivated students. It is a summer course for advanced placement high school students from all over the world, and they earn eight college credits from the University of Chicago for taking it. The students are among the brightest, most enthusiastic students I have ever taught and are exceptional workers in the field. At the end of the fieldwork portion of the course, I give each student their own specimen of *Knightia eocaena* from the quarry to keep, usually a specimen that they found themselves. This species is the most common fossil fish in the FBM, and it is the same species that was given to me by my old friend Hans long ago that motivated my switch from business school to science in college. Maybe as this little fish sits on their bookshelf at home, it will help keep their interest in paleontology alive and growing as it did for me. Mentoring these students somehow feels like a debt repaid to the spirit of the professors and colleagues who encouraged me to follow my own intellectual passion.

Although the FBM was my first and became my most important

field site, there were other parts of the world that I explored during my career. Curatorial fieldwork is sometimes opportunistic, and an opportunity involving a 100-million-year-old fossil fish locality in southern Mexico came to me during my third year as a curator in Chicago. Through the process of developing this site, I came to know many sides of Mexico and a particularly unforgettable character named Shelly Applegate.

Jim Tynsky (on tractor) and me in 1984 on Lewis Ranch. Here we just bulldozed 20 feet down through the FBM in search of the fossil-rich "18-inch layer," the paleontological motherlode.

(*Top*) After bulldozing to within a few inches of the 18-inch-layer slabs, the rest of the overlying debris is cleared away with hand shovels and leaf blowers. (*Bottom*) Then we carefully pry up the 18-inch layer in half-inch-thick to inch-thick sheets, one layer at a time.

(*Top*) As we lift up each limestone sheet, we look underneath for ridges indicating skeletons under the surface, such as the large fish I am looking at here (head facing up, covered with limestone). (*Bottom*) When first excavated, almost all of the fossils are covered with a thin layer of limestone that must be removed with special equipment (see next page).

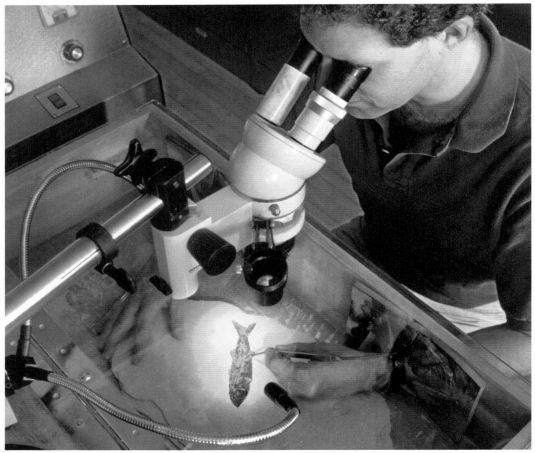

(Top) The thin layer of limestone covering the fossil is carefully removed under a microscope with fine tools. *(Bottom)* Removal of the overlying limestone reveals a beautifully fossilized fish with scales preserved in brown and bones preserved in black. (This is the piece from the bottom of previous page after fine preparation with no restoration or added color).

The FBM graphically records a dynamic 52-million-year-old subtropical ecosystem. (*Top*) Leaf with insect chew marks along its edge. (*Bottom*) Associated insects of the sort that probably made such chew marks including a stinkbug (*bottom middle*) and plant hoppers (*bottom left and right*).

There are many species of undescribed flowers and other plants preserved in the FBM. The early Eocene was a time of great evolutionary diversification among the flowering plants and the animals that fed on them and/or pollinated them.

A beautifully preserved palm frond (with several small fishes) from the FBM. Such specimens attest to the subtropical environment that existed in southwestern Wyoming 52 million years ago. I used spectacular pieces like this from my field site to help design the Fossil Lake gallery in the Field Museum's *Evolving Planet* exhibit. The exhibit is seen by over 1 million visitors annually.

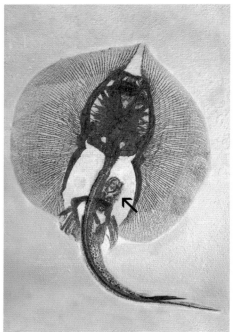
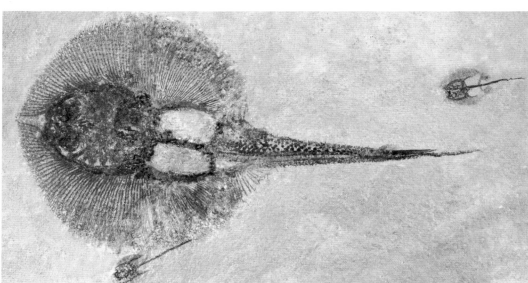

Sex and birth in the fossil, record as shown by the FBM stingray *Asterotrygon maloneyi*. (*Top left*) Mated pair with the male behind the female in copulatory position. (*Top right*) Pregnant female with embryo (complete with stinger) coiled up in her pelvic region (*arrow*). (*Bottom*) Mother and two recently born pups preserved together.

Growth series of *Diplomystus dentatus*, a primitive herring-like fish from the FBM. Specimens here range from an 18 mm individual still coiled up in the egg (*top left*) to a 24 mm hatchling (*top right*) to a 100 mm young juvenile (*middle*) to a 531 mm large adult (*bottom*). The FBM has relatively complete growth series for many different species of animals.

Dietary information is available on the FBM species because of specimens preserved with prey in their mouths and stomach contents. Here we see two specimens of the fish *Mioplosus labracoides*; one (*top*) with a fish in its jaws, and another (*bottom*) with a fish in its stomach.

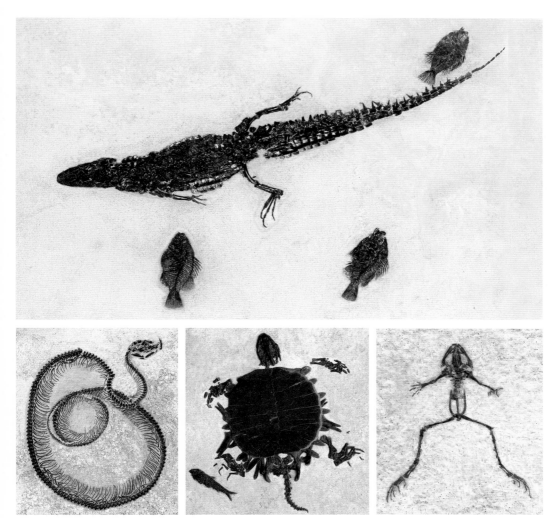

Beautifully preserved reptile and amphibian fossils from the FBM. (*Top*) caiman, *Tsoabichi greenriverensis*, with three fishes (*Cockerellites liops*); (*bottom left*) Boa constrictor-like snake, *Boavus idelmani*; (*bottom middle*) soft-shelled turtle (*Apalone heteroglypta*) and fish (*Knightia eocaena*); (*bottom right*) frog, *Aerugoamnis paulus*.

The bird fossils of the FBM are beautifully preserved skeletons, sometimes with the feathers still attached. (*Top left*) Undescribed new flightless bird; (*top right*) undescribed new land bird, with color pattern still preserved in the feathers; (*bottom left*) undescribed new bird of unknown family; (*bottom right*) *Eocypselus rowei*, a primitive bird anatomically intermediate between swifts and hummingbirds.

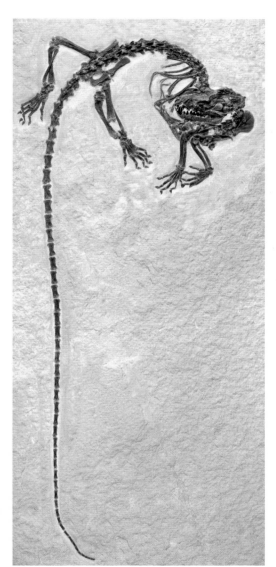
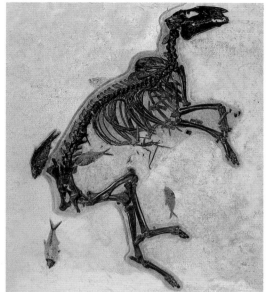

Unlike other fossil localities of similar age where fossil mammals are preserved only as fragments, the beautifully preserved mammals of the FBM are usually complete skeletons. (*Left*) Undescribed tree-climbing carnivore and earliest-known mammal with a prehensile tail; (*upper right*) three-toed horse (*Protorohippus venticolus*), which as an adult stood only 20 inches high at the shoulders; (*lower right*) the primitive bat *Onychonycteris finneyi*.

The students in my "Stones and Bones" class that I bring to Wyoming each year to prospect the FBM are some of the best fieldworkers I have ever had. Coming from all over the world, they are seriously focused on digging for fossils, and their enthusiasm is contagious even to me and my veteran crew. Shown here is one of the many motivated students from my 2011 class.

It takes a dedicated and experienced team working together to develop a productive field site, and for the last fifteen years, I have had the best. In the front row from left to right are core team members Mike Eklund (wearing dust blower on his back), Brian Morrill (holding shovel), me, Jim Holstein, and Akiko Shinya. Also shown are three recent volunteers: Jon Mitchell (behind me), Ellen Gieser (far right), and Drew Carhart (behind Ellen). At Lewis Ranch, 2015, for my forty-first field season in Wyoming.

Me in Tepexi, Mexico, in 1986 with one of the local modes of transportation. Tepexi began the expansion of my field programs to a global scale.

4

Mexico and the Hotel NSF

A few years after beginning my curatorial position at the Field Museum, I began to look for ways to expand my research operations internationally. My first success in this regard was a project with the late Shelton Pleasants Applegate (1928–2005), curator and professor at the National Autonomous University of Mexico (UNAM). Shelly had a significant impact on my early curatorial career in the mid-1980s by drawing me to Mexico for adventurous experiences that I vividly remember to this day.

Shelly was one of the most unusual scientists I have ever known. Born and raised in Richmond, Virginia, he spoke with a nasally Virginian accent and did his doctoral work at the University of Chicago. From 1962 to 1975, he had held a series of term positions at Arkansas State University, Duke University, the Smithsonian, and the Natural History Museum of Los Angeles County until finally ending up in Mexico City at UNAM in 1975, where he would spend the rest of his career. That is where I first met him. Shelly never had many publications, and of those he did have, some became well-known examples of serious mistakes in his field of research (e.g., major misidentifications of fishes). Shelly was never afraid to go out on a limb with his ideas and opinions. Large, loud, and white haired, he resembled a beardless Santa who was still wearing clothes that he had slept in the night before. He was a free spirit with a lust for life and a pragmatic attitude about how to live it. He was short on inhibition and always up for a good party. He claimed to have once been married to a "white witch," and he supplemented his modest income in Mexico with the production and sale of tarot cards. He believed in mysticism and magic, which I always found odd for a scientist. But that was Shelly. Fortunately, he was good at compartmentalizing all the facets of his richly complicated personality and never tried to foist his supernatural beliefs on anyone else.

Shelly loved his work as a scientist. In Mexico he split his time between studying living sharks on the west coast of Baja and fossil fishes in Puebla. At UNAM he was a dedicated teacher and was well regarded there as the "father of paleoichthyology" in Mexico. He built the bulk of the vertebrate fossil collection at UNAM and became its curator. Most importantly for my career, Shelly discovered and shared a new fossil locality that formed the focus of my first major National Science Foundation (NSF) research grant.

I met Shelly through my colleague David Bardack from the University of Illinois at Chicago (UIC). At the time I was still an untenured assistant curator at the Field Museum, and Dave was a full professor in the Biological Sciences Department of UIC. Shelly had been an acquaintance of Dave's and had sent him a picture of a building in southern Mexico with some highly unusual construction materials. The outer walls of the building were made of large red slabs of limestone containing ivory-colored fossil fishes. They belonged to many new and previously unknown species, and the quarry from which they had come represented a brand-new look at Earth's deep past.

The source of the fossiliferous stone was the Tlayúa Formation in a small quarry located in the poor village of Tepexi de Rodríguez, Puebla. The name Tlayúa is from the Aztec language Nahuatl, meaning "the place of obscurity." The quarry was owned and operated by Don Miguel Aranguthy and his sons Felix, Ranulfo, Faustino, Sebastian, and Benjamin. Dave and I decided to build a research project around the quarry with the aid and collaboration of Shelly and the Aranguthy family, and we submitted a seventy-page proposal to the NSF. It outlined a plan to hire the Aranguthys for a period of two years and to expand their rock quarrying operation. They would be encouraged to focus on finding fossils for us rather than simply quarrying stone slabs for construction. And for the duration of the project, Dave and I would fly to Mexico City every few months and drive to Tepexi with Shelly to examine newly excavated fossils. A year later after extensive review by the NSF and the peer scientific community, the grant was approved. I was elated. During the next two and a half years, I would acquire many lasting memories of colleagues and events in Mexico.

Dave and I arrived in Mexico City in April 1986 to begin making connections for the project. The plan was to meet Shelly the next day at UNAM to see the fossil fish collection and meet people at the university. Then, a day or two later, we would drive to Tepexi and see the operation there. Dave and I checked into a high-rise hotel in the central part of the city. At dinner Dave told me a few amusing stories about Shelly, and we were both trying to learn a few critical phrases in Spanish in case we needed them, such as "*¿Cuántos pesos?*" (How many pesos?), "*El agua embotellada, por favor*" (Bottled water, please), and "*¿Dondé está el baño?*" (Where is the bathroom?). After dinner I went to my room on the fifth floor.

Sleep came quickly that night. I clearly remember dreaming about a carnival ride and slowly coming out of that dream at about 1:00 a.m. to feel the bed violently moving back and forth. There was dust in the air, and small pieces of the ceiling were falling down. I leapt out onto the balcony of my room just in time to see a bright flash light up the night sky. Alarms and sirens were going off everywhere. In my jump-started wakeful state, my first thought was that a nuclear bomb had gone off. *What luck,* I thought, *my first freakin' day in Mexico and I'm going to die here.* Stunned, I stood there looking out over the city as things finally stopped shaking. Then I slowly came to the realization that this was not a bomb. The flashes were lightning, and the shaking was from an earthquake. As we found out later, Mexico City had just experienced a large earthquake. The next morning as Dave and I stood outside in front of our hotel, we could see a large crack going through the outer wall of our hotel, and in the distance another building had collapsed. The phone lines were all down in the city, and we had no way to call home to let people know we were all right. It was then that we decided to move to Shelly's house, which was safely located on firm bedrock in the mountains overlooking the city. We found a driver whom we paid to take us to up there. When we arrived, Shelly was just rolling out of bed. He had slept right through the quake, which had been much milder in the mountains. He greeted us loudly and enthusiastically, inviting us to move in for as long as we wanted. And that was when we began to form a close friendship with our colleague Shelly.

The house was on land that Shelly did not legally own. He was an immigrant squatter of sorts, and his neighborhood was a bit like the Wild West. Electrical wires from the houses in Shelly's neighborhood were jury-rigged to splice in to the closest public power lines and poach off of the grid. Shelly's house was always a work in progress. It grew a bit each year as he found the money or connections to put on another wall or two. It was made almost entirely of concrete, brick, and cinder block. The structure had exposed sections of iron reinforcement bars (rebar) extending from the last walls to have been installed, as though waiting eagerly for the next construction phase to begin. The house and tiny grassless front yard were surrounded by an eight-foot-high concrete-and-brick wall with sharply edged broken glass set into mortar along the top of it. There were big iron doors set into the wall guarding the entrance to his driveway, and Dave and I would later occasionally refer to the place as Fort Shelly. Shelly might not have been so concerned about the house's property title, but he was clearly worried about security. Although the outer walls presented a cold, hard appearance, the inside of Shelly's house was a different story altogether. It was a home of warmth and celebration. Walls and floors were bright with colorful tiles in places, and the center part of the house was laid out with an open floor area for parties. The central area also had large windows to let in the sunlight or moonlight, depending on the time of day. His private bathroom had a very large tub lined with small yellow ceramic tiles, big enough for two people of even Shelly's considerable size.

Shelly's house was only a day's stopover during our trip to Mexico. The ultimate destination was the small village of Tepexi, and Dave and I were excited about the prospect of seeing the fossil quarry for the first time. Shelly had an eighteen-year-old Ford truck that served as our transportation. It appeared to be destined for semi-immortality through mummification. Trucks in southern Mexico were a precious commodity at the time, and they would be repaired and maintained well beyond what their expected life span would have been in the United States. Shelly's truck looked particularly precarious and was referred to by his colleagues at UNAM as *El espantomóvil* (the ghost mobile). Dave and I could only hope for the best when Shelly used the ghost mobile to ferry us across many miles of isolated Mexican desert on the way to Tepexi. We were also dependent on Shelly to be our translator because neither of us could understand much Spanish. This was a leap of faith that Dave and I would occasionally regret having made. Shelly claimed to have learned all the Spanish he needed from Mexican soap operas on television, which he watched regularly. Although he had been living in Mexico for many years when I first met him, his Mexican colleagues said he still spoke Spanglish, that unpredictable mix of English and Spanish, overlaid with his heavy Virginian accent. Simple conversations between Shelly and the residents of Mexico could sometimes take a long time and often resembled an enthusiastic game of charades. But his Mexican colleagues joked rather than complained about it, and they were very patient. And fortunately for us, Shelly's translations turned out to be sufficient most of the time.

Tepexi was 140 miles southeast of Mexico City in the southern part of Puebla, and to get there was a three-hour trip in Shelly's aging vehicle. As we departed Mexico City, Shelly warned us that we would go through a couple of dangerous areas on the way through the desert. We could possibly run into drug runners who might try to flag us down or stop us. He told us flat out not to try to prevent him from running these people down with his truck if they tried to physically block our way. He was dead serious. The ghost mobile versus drug dealers? Shelly went on to explain that he had already experienced two close calls with armed bandits and some mahogany tree poachers on previous driving trips, and he had no wish to invite any third-time's-the-charm karma. Dave and I were starting to wonder what we had gotten ourselves into, but there was no going back now. So as we set out for Puebla, Dave and I put ourselves in the hands of our new colleague, Shelly.

We made it to our destination without any problems. The area was a sunny desert environment, hot and dry. Surrounding the village, there were majestically beautiful forests of green columnar cactus standing in contrast against a red desert landscape. Many of these cacti were 20 to 30 feet tall, with stemless, daisy-like flowers sprouting from their trunks. Brightly colored hummingbirds darted back and forth among the cactus blooms. The Aztecs believed that hummingbirds were reincarnated warriors who had been killed in battle. Coming into the Aranguthy compound was like going back in time. The area was a beautiful but particularly poor region of Mexico that had been that way since the time of the Aztecs. The family had several primitive-looking buildings on a few acres of land. The houses were made largely of stone, cinder block, and mortar with thatched roofs. Barns and storage buildings seemed to be made of sticks and logs. The houses were without running water, which had to be carried some distance from a stream in the valley by donkey (a main method of transportation) or by people (mostly women and children). In those buildings that had electricity, electrical appliances consisted mainly of bare lightbulbs hanging from the ceilings. Dave, Shelly, and I all slept in cots in one 8-by-10-foot room with a stone floor, cinder-block walls, and a metal roof. At night the little room echoed with the sound of Shelly's thunderous snoring.

Many of the children we saw on our first day standing outside were barefoot. It was clearly a marginal existence for many people of the area, but they seemed happy and content. Many of the residents were pure-blooded indigenous Mesoamerican people who still spoke in pre-Spanish languages such as the Aztecs' Nahuatl or the ancient Mixtec. The region was literally awash with antiquity. During heavy rainstorms, pre-Columbian artifacts more than a thousand years old would wash out of the hills. Some of the nicer-looking relics (small bowls or pieces of statues) had been picked up by the Aranguthys and were being used as ornamental decorations in their home. Shelly found a pre-Columbian whistle after a storm. He cleaned it up, and after it dried, it still worked! Priceless artifacts washed out of the hills on a regular basis, and Shelly said the museums could not collect them all for lack of museum storage space. It was a humbling experience for a museum curator such as me. One day I almost stepped on an obsidian knife blade that had washed out of the dirt. I picked it up and stared at it in amazement. Who knows what it had once been used for? Cutting meat? Human sacrifice? The blade would have been an exhibit piece back in Chicago, but I gave it to Shelly and he put it back on the ground. It was highly illegal and strictly forbidden to take any pre-Columbian artifacts out of Mexico.

Mixed with the general atmosphere of cultural antiquity, we could see superficial traces of modernization. The Aranguthys had an old refrigerator, for example, but the refrigeration part of the appliance did not work. It was used simply as an insulated storage cabinet for water, vegetables, and a few bottles of warm Dos Equis beer. They had a box of Purina dog chow for the family dog and for the goats on occasion. And there was a Dos Equis beer truck that came to town and made deliveries, much like the milk truck I remember as a child growing up in Minneapolis. And of course there was Shelly's ghost mobile and a couple other old trucks around the property. Other than that, there was not much to be seen of twentieth-century technology in this part of Mexico. Dave and I were both stunned at how happy the family seemed in an environment lacking so much of what Dave and I had considered to be necessities of life. There was magnificence in the simplicity of life here.

Our plan for the NSF project was to pay the Aranguthy family a full working wage to focus on mining the quarry specifically for fossils instead of building stone. We had decided that $18,000 per year would be a fair wage. Then we would come down to check on the progress every few months over the next two years. As it turned out, our estimates of wages for the Tepexi region were well on the generous side for the mid-1980s because it was tied to the U.S. dollar. The Mexican economy at the time was on the verge of collapse, the value of the peso was sinking like a rock, and the value of the dollar in Mexico was rocketing. By the end of the first six months, the amount we were paying the Aranguthy family was about five times what they had previously been making as stone quarriers. Their local standard of living increased sharply. By the second time we went down to Tepexi, all the children were wearing shoes. The family had bought a working refrigerator and erected a new building, and Felix Aranguthy was running for local political office. He had become a local celebrity partly because of attracting foreign visitors and money into the area. I started to wonder if we had not inadvertently upset the normal course of cultural development here (probably a subconscious flashback to my adolescent years of watching the TV show *Star Trek* and its ever-present "prime directive" prohibiting interference with indigenous cultural development).

After a few trips to see the progress in Tepexi, Dave and I were becoming popular visitors. Unfortunately, we were totally dependent upon Shelly's Spanglish translations of what was going on. Nobody we dealt with in Tepexi, and I mean *nobody*, spoke English. In retrospect, it is amazing (and retrospectively frightening) to me how many ways we were dependent on Shelly. Shelly wasn't always accurate in his translations, and the locals had trouble understanding him. On rare occasions we would see angry faces of frustration when communicating through Shelly. The scariest of those times was at one of several fiestas that were thrown in our honor when we came down during the second year. It was a big celebration. The project had been going well, and there was even a new small building of sticks and logs that the Aranguthys had built to store the NSF material. There was a sign over the door saying something to the effect of Hotel NSF. Because Felix Aranguthy was interested in local politics, he had invited the chief of the military police, the chief of the secret police, and other high-ranking police and military officials from the region. When they showed up, some were wearing guns and they came with their girlfriends. One of them even wore a small bandolier of bullets, like in an old Emiliano Zapata movie. We were at first both fascinated and amused. But later that night after having a few too many tequilas for clear judgment, Shelly made a pass at one of the armed gentlemen's girlfriends. The mood in the room changed instantly. As Shelly rapidly tried to explain himself in Spanglish, the faces got angrier and the yelling got louder. One of the armed partygoers even put his hand on his sidearm. I leaned over to Dave and rhetorically asked, "What the hell do we do now?" Dave just sat there in a silent state of wide-eyed, nervous awe watching the events unfold. Then all of a sudden, Shelly quietly said something to the agitated crowd that Dave and I could not understand (and to this day we do not know what it was); and like a magic switch had been thrown, everyone was laughing once more and slapping each other on the back. Like so many times before, Mexico had shaken us for a moment, but it all came out fine in the end.

There was one more event during the second year that was particularly memorable. It was on a return flight from Mexico. After we boarded the plane at Mexico City International, it taxied for a while and finally stopped near the outskirts of the tarmac, where it sat for several hours. Finally, as we looked out the window, a truck filled with uniformed military troops drove up. The men filed out of the trucks and surrounded the plane. We were all asked to exit the plane down ladders to claim our luggage, which had been unloaded on the runway. When all passengers had claimed their pieces of luggage, there was one small bag remaining. Then another truck with a thick wire mesh trailer drove out and two men in flak jackets loaded the remaining bag on a stretcher. They put the stretcher with the mystery bag in the trailer and drove off. After that, we reboarded the plane and it took off with nary a word from the pilot. The whole ordeal had kept us on the runway for more than five hours. Dave and I figured it must have been a bomb threat (or even a real bomb), but we could only speculate. This event would have made front-page news in the United States. But here in Mexico, it went unmentioned as far as we were ever able to tell. It was all part of the charm and exotic nature of Mexican culture in the 1980s, and an important lesson about curatorial fieldwork in the developing world. You must network closely with local colleagues, be adaptable, and expect the unexpected.

During the two-and-a-half-year project, hundreds of fossils were excavated from the Tepexi quarry. Most of them were fishes, but there were also plants, invertebrates, turtles, lizards, crocodiles, and a pterosaur. We determined that the deposit was just over 100 million years old (from the Albian Stage of the Cretaceous). There were many new species to be described, including the oldest-known species of the herring family (fortuitous for me, having done my PhD thesis on herrings and their close relatives). I used some of the Tepexi material for a large research project on bowfin fishes that I later published with my colleague Willy Bemis, who is featured in chapter 5. This species, *Pachyamia mexicana*, was particularly interesting in that its closest relative was another fossil bowfin found 20 miles north of Jerusalem, in the Middle East. The relationship between the Middle Eastern species and the Mexican species reflect the different positions of the continents 100 million years ago, when there was no deep Atlantic Ocean separating North America from the Middle East. The Tepexi locality provided the first detailed biodiversity survey of the Middle American region during the Albian time period and documents an important piece of Earth history.

Shelly continued to oversee the collecting of fossils from the Aranguthy quarry after the NSF project ended in 1989, up until his death in 2005. In the end, the collection amounted to over 6,000 fossils. The majority of specimens excavated went to the paleontological collections of UNAM in Mexico City, and a small collection of specimens went to start a small local museum in Tepexi de Rodríguez that was directed by Felix Aranguthy called the Museum Pie de Vaca. Lastly, a representative sample of the fossils was incorporated into the collections of the Field Museum as part of the original collaborative agreement between the Field Museum, the Aranguthy family, UNAM, and government officials of Puebla, Mexico. The Field Museum in Chicago is the only institution outside of Mexico to have a legally obtained a collection of fossils from the Tlayúa quarry, adding one more unique element to its research collection.

Most of the Chicago specimens will be studied by someone other than me because they fall into families of fishes outside my own specialty and interest, but the addition of the Tlayúa Formation fossils to the Field Museum's collection is an example of how curators of major natural history museums build diverse collections for research on a global scale. What is not studied by me will one day be critical to someone else's future research project. And Shelly? His collection-building activity at UNAM helped create a paleontology program in Mexico that is still thriving today, inserting Mexico more firmly into the international network of natural history researchers. His students became some of Mexico's first fish paleontologists. And maybe there was something to that mystical nature of his, because his colleagues tell me the spirit of Shelly remains in Tepexi, and in the collection halls of UNAM.

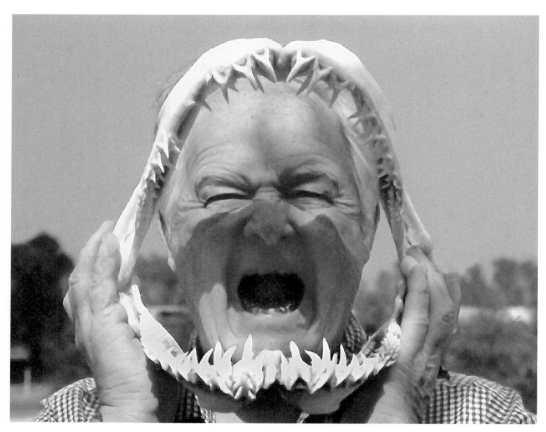

One of my earliest international expedition colleagues, Shelton P. Applegate, in Baja, Mexico, with a shark jaw. He was key in helping me get my first major NSF research grant and in drawing me to Mexico.

"Fort Shelly," home of Shelly Applegate, nestled high in the safe mountain bedrock surrounding Mexico City in 1986. It was a welcome refuge for Dave and me after being shaken the night before by a powerful earthquake.

Shelly's "ghost mobile" (*El espantomóvil*) with Shelly (*left*) in 1986. This truck was our main source of transportation in Puebla, Mexico. It was first-class wing-and-a-prayer travel, but Shelly wouldn't hear of our renting a vehicle.

Some of the Aranguthy family men of Colonia Morelos with David Bardack and me in 1986. I am crouched down on one knee in the lower left of the picture with Dave (to my left), and Aranguthy patriarch Don Miguel (to Dave's left behind the goat). In the back row from left to right are three of Don Miguel's sons, Faustino, Sebastian, and Benjamin. Also present is an unidentified quarry worker (in the green and yellow baseball cap) and a neighbor (to Don Miguel's left).

Some of the Aranguthy children in front of their house. Life in Colonia Morelos seemed very peaceful and uncomplicated.

(*Top*) Shelly and me walking through the part of the Aranguthy quarry expanded with our grant from the National Science Foundation. (*Bottom left*) Aranguthy family members working in the quarry and splitting slabs of 100-million-year-old limestone. (*Bottom right*) Split slab revealing a curled-up needle-snouted fish, *Belonostomus*.

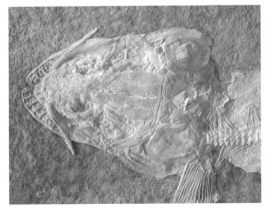

Aranguthy quarry fossils are preserved as ivory-colored bones in pink or red limestone. (*Top*) New species from the extinct family Macrosemiidae; (*bottom left*) oldest-known member of the herring family; (*bottom right*) head region (squashed in top view) of a primitive bowfin, *Pachyamia mexicana*, named and described with my colleague Willy Bemis in 1998. (More on Willy in the next chapter.)

The tall cactus forest surrounding the Aranguthy quarry, with some cacti reaching 20 feet or more in height. It remains as one of my strongest memories of the Colonia Morelos desert.

Willy Bemis in his lab at the University of Massachusetts in 2002 with skeletons from the Alabama Deep Sea Fishing Rodeo. Willy became my most important research collaborator over the years as well as a close friend with whom I shared many adventures.

5

Willy, Radioactive Rayfins, and the Fish Rodeo

Ahh, Willy . . . What do you say about a colleague who over the years comes to feel like the brother you never had? Willy Bemis fits well into my eclectic assemblage of people most important to my development as a curator. My work with Willy expanded my exposure to the travel requirements of curators interested in doing research on a global scale. We visited museums and fossil localities all over the world in a collaborative research program with each other that lifted my curatorial career to the next level. And over the course of this fifteen-year project, we became close personal friends.

By the late 1980s my project in Mexico had ended, and I was looking to build a long-term collaborative research program on fish anatomy and evolution. Willy was a professor of biology at the University of Massachusetts whom I had known for years. He had an encyclopedic knowledge of fish embryology and soft tissue anatomy, which was a prefect integrative match for my experience with skeletal anatomy and paleontology. In the summer of 1987, we met at the museum and agreed to build a collaborative research program on rayfin fishes. Rayfins (scientific name Actinopterygii) include most of the 35,000 species of fishes living today, and they have a fossil record extending back over 400 million years. It is the most successful group of vertebrate animals on Earth. We would focus on several groups near the base of the rayfin evolutionary tree that still have a few living species, including sturgeons, paddlefishes, gars, and bowfins. We would study all known species within these groups through time (fossil and living) and space (all parts of the world). Our study would ultimately help other ichthyologists working on rayfin groups higher up in the evolutionary tree by determining the significance of many rayfin anatomical characteristics at the base of the tree. It would test new computer programs designed to calculate cladograms in the process. Together, Willy and I submitted several proposals to the National Science Foundation, and we were very successful. Between 1988 and 2000, we received five major NSF grants totaling over $1 million for our work on rayfins. The National Science Foundation, more than any other funding agency, made the type of basic research that Willy and I did possible. The project took us all over the world, including twelve different countries and a dozen different states in this country.

In May 1994 we set out for Moscow and Jerusalem. It was only a few years after the fall of the Soviet Union, so things were still in transition there, creating great problems with public services and infrastructure. We spoke with Russian colleagues before leaving to see what we would need for a week in Moscow. We were told to bring plenty of U.S. dollars in small denominations and unmarked bills. At that time many Russians did not like taking rubles (the Russian currency); they were suspicious of any U.S. bills that had writing or other markings on them (thinking they might be traced for some reason or, worse yet, counterfeit), and nobody in Moscow gave change. This would require us to carry large sums of money, mostly clean new bills in denominations of $5, $10, and $20, stuffed into money belts. It reminded me of going to Mexico in the mid-1980s when inflation was rampant but the highest denomination of currency had not been increased accordingly. Willy and I joked that our fat money belts made us rich, but in reality the large bundles of cash amounted only to survival money for life in a very challenging economic environment. We were also told by a Russian colleague visiting the Field Museum that we should bring coffee and chocolate as gifts for the Russian scientists, and to bring bottled water to avoid bacterial problems with the city's tap water in the hotel where we would be staying. We bought a large red suitcase for $5 at a Salvation Army store in Chicago and filled it to capacity with the gifts and bottles of water to take on our journey.

After an eleven-hour flight, we arrived at the Moscow airport, where we were met by our host Eugenia K. Sytchevskaya, who had generously offered to drive us to our hotel. She was the curator of fossil fishes at the Paleontological Institute of the Russian Academy of Sciences (PIN), one of the world's largest paleontological institutes with well over a hundred research scientists. It included a large museum with exhibits and huge collections from all over the former Soviet Union. On our way to the hotel, Eugenia gave us a few words of wisdom for life in post-Soviet Moscow. She told us to avoid the downtown area because since the fall of the Soviet Union and the onset of rapid privatization, it had become a mecca for the Russian Mafia. The political transition had brought on a great experiment in jump-started capitalism. As we drove through part of downtown Moscow, it looked like a mini Las Vegas, with more than seventy different brightly lit gambling casinos and a host of expensive hotels, shops, and restaurants. Large billboards appeared everywhere for local investment banks offering 50 to 100 percent annual interest on savings accounts (which in later years would prove to be financially fatal to most of those institutions and their customers). The transition toward capitalism—or "privatization" as it was called in the 1990s—was in an uncontrolled full swing. The major financial benefits of privatization went disproportionately to select individuals, resulting in Moscow eventually having the largest concentration of billionaires for any city in the world (per *Forbes*' 2011 annual list). In contrast, as we found out later, Eugenia and her colleagues at the PIN were facing a combination of rising inflation and decreased public services that was straining their budget for day-to-day existence.

Eugenia dropped Willy and me off at the entrance to the Hotel Uzkoe. As opposed to the hotels in downtown Moscow that were hundreds of dollars per night even in the mid-1990s, the Uzkoe was less than a mile from the museum and was only about $50 per night. As we checked into the hotel, we were told that there was no hot water in the hotel or surrounding area. There had been a breakdown in the central pumping station for hot water. In much of Moscow, it was rare to have individual hot-water heaters in the homes and hotels. For the most part, hot water and heat was piped underground from a monstrously inefficient centralized heating facility in the city. Moscow's heating system was reported to be annually consuming as much natural gas as the entire country of France. Heat was treated as a public utility much like electricity and gas, and its delivery system was an infrastructural vestige of Soviet-era Big Brother control. It seemed incredible to us having grown up in a country where every house has its own individual water heater and furnace. We were also warned politely that we should stay out of the bar and lobby area between 11 p.m. and 2 a.m. because it was reserved for a group from downtown Moscow affiliated with the Russian Mafia who could sometimes be unpredictable with foreigners who did not speak Russian. We had a Spartan dinner in the hotel restaurant and retreated to our rooms for the night so we could get an early start the next morning.

The first day at the PIN was another cultural adjustment. The building was large, with an exhibit area of fossils ranging from snails to dinosaurs. It was darker than we were used to in U.S. museums, and as with our hotel it reflected the more frugal side of life in Moscow. The museum's public restrooms had no toilet paper. Instead, sitting beside a toilet with no toilet seat was a board with a large nail in it and impaled on the nail were dozens of 5-by-5-inch squares cut out of newspaper. Many other basic supplies were in short supply as well. In the collection rooms, there were not enough lightbulbs for all the light sockets. As Willy and I ventured down some of the dark collection aisles, we would have to unscrew lightbulbs behind us to screw into empty sockets in the direction we were going. The early and mid-1990s were challenging times of austerity in Russian academic institutions. What was most important to us, though, was that the PIN had huge collections of fossils that few scientists from the Western world had ever seen, and the scientists of the PIN gave us free access to whatever we wanted to study.

The PIN had some particularly interesting fossil rayfins from 60-million-year-old rocks from Mongolia. Although they had already been named and briefly described by our colleague Eugenia, there were still undescribed parts of anatomy that we needed to see for our research. The specimens were beautiful, but there was one complication. They were radioactive! They were kept in special lead-lined cases because they contained high amounts of radium. Radioactive fossils in an institution that currently couldn't afford toilet paper and lightbulbs. We nevertheless spent a couple of days looking closely at the material, wondering in the back of our minds what it might be doing to our future offspring.

We spent the rest of the week taking photographs and making drawings of dozens of fossils. It was still the pre-digital age for photography, so Willy and I had brought a small mountain of photographic equipment in order to take hundreds of pictures using 4-by-5-inch negatives. We developed the negatives there to make sure that the images were adequate, because we knew we wouldn't be coming back to Moscow anytime soon. The information and images we recorded from the radioactive species and a number of other fossils enriched our cladogram of rayfin fishes. It was a highly productive trip.

From Moscow we flew to Israel to study fossil fishes that were unique to a locality there. The flight to Tel Aviv was long, because it was routed through Frankfurt, and the German and Israeli security forces did not appear to be particularly trusting souls for passengers arriving from Moscow. There were guards with machine guns in black uniforms everywhere. I was chauffeured into a large indoor tent and thoroughly frisked, after which every piece of my luggage was opened and examined. After Willy and I made it through the long entry process into Israel, we came out into the bright light of day and everything changed. The first thing that struck me was the warmth I immediately felt after emerging from the airport. Moscow had been cold, damp, and gray. In contrast, the bright sunlit sky of Tel Aviv was a deep blue and the air was clean, warm, and dry. I immediately felt uplifted and energized.

We caught a cab to Jerusalem. The farther we went, the better things looked. Date palms and other fruit trees seemed to be everywhere, as were lush groves of olive trees. Our destination was the Paradise Hotel, which seemed aptly named after our week in Moscow. Fresh fruits, vegetables, meats, cheeses, and fine wines were plentiful in the restaurants and stores. The hotels were spotless and brightly lit. The next morning we went to the Hebrew University to see Eitan Tchernov, curator of the paleontology collection for the university, and fossil fish expert Yael Chalifa. We had planned to collect fossils in a Cretaceous quarry located in Ramallah, part of the West Bank region that had been captured by Israel during the Six-Day War in 1967 and that contained controversial Israeli settlements of questionable international legality. The quarry had produced a particularly interesting species that we needed to see for our study. We were excited about the trip, seeing the quarry, and the possibility of finding more material. But on the evening before we were to visit the quarry, there was a shooting followed by a mob-led stoning incident, and our hosts told us it would now be too dangerous to go there. Willy and I would have to be satisfied with looking at specimens in the collection of the Hebrew University's museum. Fortunately for us, these specimens turned out to provide much of the information that we were looking for.

Near the end of our trip, Yael took us to see a fossil collector near Ramallah who she thought might also have material that would interest us. On the way there, we had a flat tire. After we pulled over to the side of the road to fix it, another car pulled in behind us. Yael became extremely nervous and asked us to stay in the car while she went to speak with the people in the other car. She seemed emphatic about our staying put, so we complied. She told us that she recognized the license plate as one from a Palestinian area that might mean trouble. She got out of our car and went back to the car behind us. As Willy and I watched through the back window, she exchanged a few words with the driver, who had also left his car. They spoke to each other, looking tense and guarded for what seemed like a long time but was in fact only a few minutes. Then Yael came back to the car and said everything was all right. But Yael's nervousness did not subside until the other car drove off. We then got out of the car and changed the tire and went on our way. This incident and comments from other colleagues there showed a constant state of tension underlying day-to-day life in Israel. Once we arrived at the collector's house, we were greeted and invited into his beautiful home. His house was in the occupied territory region of the West Bank. It looked out over miles of undeveloped desert-like hills that reminded me of the desert buttes of Wyoming where I do fieldwork each summer. The collector had many fine fossil specimens in his collection. He had bought them from quarriers in Ramallah over the years and had amassed a collection of fossil fishes almost as great as that of the Hebrew University. We studied what he had there for an hour, and when it was time to go, he let us borrow some of his material for further study in Chicago. It was one of many places around the world where we had to make a personal appearance to hand-carry material in order to borrow it for our studies in Chicago. As we left the collector's house, we said our good-byes and thanked him for his hospitality.

Like the Moscow leg of this trip, the Middle East stop was a success. We had plenty of material to take back to Chicago for months of preparation and study. The Ramallah fossils provided Willy and me with anatomical information to help resolve the rayfin fish tree. They also provided further evidence of the 100-million-year-old connection between southern Mexico and the Middle East that I had detected while working in Mexico. And Willy and I were building an international reputation of trustworthiness among people and institutions from which we needed to borrow specimens.

Early in the following year, our project took us to Kitakyushu, Japan. There were fossils that we needed to see there under the care of curator Yoshitaka Yabumoto (Yoshi) of the Kitakyushu Museum of Natural History and Human History. I was particularly interested in fossil rayfin fishes they had collected from Asia and South America. Previous work I had done on fossil rayfins from Wyoming indicated that western North America had a close relationship to eastern Asia and Australia 52 million years ago. This paralleled findings of paleobotanists who saw 48- to 52-million-year-old plants of western North America having a closer affinity to plants from Asia than to those from Europe.[1] This pattern of relationship between western North America, eastern Asia, and Australia supports a controversial hypothesis by geophysicists Amos Nur and Zvi Ben-Avraham about a lost continent they called Pacifica.

Pacifica was a large continental region that began to break up just over 200 million years ago (see page 101). Large sections of Pacifica separated from what is today the east coast of Australia and slowly migrated across the Pacific, riding on the natural spreading ridges of the Pacific seafloor. Tens of millions of years later, the "drifting" pieces of Pacifica collided with continental margins of North America, South America, and Asia. Evidence of these collisions today includes pieces of exotic continental terrain found today in Pacific coastal regions. One such piece is called Wrangellia, a block that collided with the Pacific coast of North America during the Late Cretaceous and contributed to the intensive mountain-building activity of the region. Also, the Pacific seafloor is still measurably spreading today at the rate of several inches per year. The lost subcontinent of Pacifica could explain the close relationship among the freshwater fishes and fossil plants of Australia, Japan, China, and western North America: they all descended from species originally endemic to Pacifica. Because the processes of fragmentation, trans-oceanic movement, and continental collision happened very slowly (only a few inches of movement per year), plant and animal species could have been safely transported. The strong trans-Pacific pattern between western North America and Asia eventually disappeared during Oligocene time, when geologic evidence indicates that the seaway dividing eastern and western North America had dried up and the Rockies were eroded to their lowest point. The loss of these barriers allowed species with transatlantic relationships from eastern North America to invade western North America, resulting in today's mixed biota.

The Pacifica hypothesis is an example of how integrative research involving different fields of science can point to larger questions and issues (in this case combining studies of fossil fishes, fossil plants, and geophysics). The integrative way that the Pacifica theory originally developed from circum-Pacific relationships and geological processes was not so different from the way that continental drift theory itself developed from circum-Atlantic patterns of relationship, geological processes of oceanic crust movement, and today's continental land shapes. The Pacifica theory of plant and animal dispersal is controversial within the geological community as a possible transport mechanism for freshwater organisms (e.g., some geologists contend that these continental fragments would have been under seawater for part of their journey), but controversy has always been part of scientific progress. Even the general notion of moving continents, which is universally accepted today among geologists, was hotly debated for decades. The famous American Museum of Natural History paleontologist George Gaylord Simpson, discussed earlier as a leading critic of cladistics (chapter 1), was also a leading critic of continental drift for over twenty years. Science is an ever-evolving process.

In our off-hours in Kitakyushu, Willy and I explored the local culture and made friends with Yoshi and other people there who had made our trip successful. Much of this socializing involved dining. Kitakyushu is on the coast of the Hibiki-nada Sea, and in this region almost all available protein came from the sea. Their most celebrated and notorious delicacy was fugu (pufferfish). We were there in February during the Fugu Festival, and there were billboards and posters advertising the cherished fish everywhere. There were fugu balloons, fugu kites, fugu lamps, fugu flags, fugu cars, and fugu trucks. There was even a fugu dirigible flying around in the sky. It was amazing to see the love that the population had for a fish that if not prepared correctly can kill you in a matter of hours. Certain parts of the fish contain a poison that paralyzes the muscles of the victim while they are still conscious; the victim is unable to breathe and dies of asphyxiation. The poison is 1,200 times stronger than cyanide. There is no known readily available antidote, and people in Japan die each year from eating improperly prepared fugu. But because it was the Fugu Festival in Kitakyushu, everyone we met there who took us out to dinner insisted on buying us fugu. We tempted fate three times over the six days we were there. One lesson I learned over my years of travel as a curator: building trust and valuable networks across geographic and cultural boundaries involves respecting your host's traditions. Sometimes that means having fine wine and cheese in France or bangers and mash in England. Other times that means dining on baked corn blight and barbequed *chapulines* (grasshoppers) in southern Mexico or on sea urchin and raw fugu in Japan.

Through our travels, Willy and I developed a strong network of international colleagues. In addition to Russia, Israel, and Japan, the rayfin fish project also took us to Italy, Spain, Germany, Belgium, Austria, France, England, Canada, and Mexico. Domestically, it took us to Utah, Colorado, Kansas, Alabama, California, Wyoming, Pennsylvania, Massachusetts, Connecticut, New York, and Washington, DC. Willy's unpretentious talent for friendly persuasion often came in handy during our heavy travel schedule. Some of our international flights would last ten or more hours. As we would inevitably come in at the last minute to the counter at the airport boarding gate to claim our seats in coach class, Willy would strike up a charming conversation with the woman behind the counter. He would eventually say something to the effect of "My friend and I have been traveling for weeks now, and it is still such a long way home. Is there any way you could see yourself bumping us up to business class?" The first time I heard this, I almost laughed out loud, thinking, *This guy is nuts!* I stopped doubting Willy the second time the ploy actually worked. There is nothing like getting bumped to business or first class for free on a long transoceanic flight.

Willy's natural abilities of persuasion extended to other facets of his professional life, especially when it came to acquiring fish specimens for our work on rayfins. He collected dead fishes from aquarium stores, beaches, fishermen, and seines. He raised fishes in tanks, ponds, and even in an in-ground swimming pool he had in his backyard (much to the dismay of his wife). Our research project required hundreds of rayfin skeletons. The most successful collecting venture of Willy's was the Alabama Deep Sea Fishing Rodeo (ADSFR), for which I joined him in 2002.

The ADSFR was founded in 1929 and is based on a small barrier island off the coast of Alabama called Dauphin Island. It is the largest fishing tournament in the world, according to the *Guinness Book of World Records*. The three-day contest involves more than 3,000 anglers from over twenty-four states who fish 45,000 square miles of the Gulf of Mexico. Nearly a thousand participating boats pull into and out of the island port throughout the day, dropping off their potential trophy winners. Standing on the shore during the rodeo, you can see a line of boats waiting to dock that stretches out as far as the eye can see. The serious contending boats, mostly 20- to 50-footers, have names like *In Too Deep*, *Play N Hookie*, and *Dirty White Boys*. The tournament awards over $400,000 in prizes for thirty different categories of competition, from biggest barracuda to biggest wahoo (yes, that is a kind of fish). To top it all off, there are 75,000 spectators.

When I arrived on the island in mid-July, the sky was overcast and there was a thick, humid breeze coming off the ocean. The smell of the sea and fish was everywhere. Willy met me in his rodeo uniform, consisting of shorts, sandals, and a floral print Hawaiian shirt. He took me to the dorm-like building where the scientists and graduate students lived and gave me a quick tour of the operation. Willy had created a dissection room in a screened-in breezeway between two buildings of the nearby Dauphin Island Sea Lab. The Sea Lab had been converted from an old army post named Fort Gaines, best known for its role in the Battle of Mobile Bay during the American Civil War. This was the battle in which Admiral Farragut supposedly uttered the legendary phrase "Damn the torpedoes—full speed ahead!" In a way, the phrase also summed up Willy's determination when he faced the many challenging parameters of the fishing rodeo. The makeshift dissecting lab had a crushed-shell floor and large exhaust fans to blow the smell of dead fishes out of the room. Willy had set up two large wooden tables in the middle of the room that could hold fishes weighing from a few pounds to a few hundred. Just outside the dissection lab, there was an enormous pit dug for the disposal of a ton or more of fish guts that would accumulate over the next week during preparation of the skeletons for shipment to UMass and the Field Museum. Willy had transformed an empty space into an effective processing center that would convert tons of would-be trophy fishes into museum research treasures over the next several days. We would both collect fishes for the UMass and Field Museum collections, some of which we could potentially use in our rayfin fish project.

At times it was mayhem. I was nearly run over by a fast-moving wheelbarrow full of 40-pound tuna one day. In spite of the circus-like atmosphere, the tournament is a boon to fish research. It provides valuable specimens to many museums and universities, sponsors several college scholarships, helps support the Dauphin Island Sea Lab, and generates over $100,000 a year for the University of South Alabama Department of Marine Biology. From 1997 to 2005, Willy manned the booth for the "Most Unusual Fish" competition, established years earlier by my former PhD advisor Gary Nelson. Gary had used it to add thousands of fish skeletons to the American Museum's collection over the years, and when he retired and moved to Australia, he passed the opportunity down to Willy, who eagerly stepped into the role. The prize for the most unusual fish was $200, but in order to enter the contest you had to donate your fish to science (i.e., Willy). This was a very effective way to harness the efforts of nearly a thousand fishing boats in the tournament to collect a diverse array of fishes for museum collections, and another example of the citizen-science approach to the collection of specimens and data for research. During the years Willy did this, he was able to collect over 250 different species of fishes by engaging the efforts of hundreds of fishermen and fisherwomen. Willy offered an additional service to the rodeo fishermen who brought in very large valuable fishes like giant groupers, tuna, and marlin. He would stand there at the docks for hours in his shorts, sandals, and shirt covered with fish slime and blood, watching the boats pull in with their potential trophy winners. When he saw someone come in with a particularly spectacular behemoth, he would approach them, offering an expert filleting service. In exchange for their donating the skeleton to science, Willy and his team (including me in 2002) would neatly and thoroughly fillet the meat off the bones for the owners. This was followed by an assurance that the donated fish skeleton would be an important contribution to scientific research. People would rarely refuse his persuasive talents. The filleting service was an effective way to collect skeletons of ichthyological giants of the sort that were normally out of reach for museums.

One fringe benefit of the rodeo was an abundance of surf selections to be had for culinary purposes. Many of the fishing contestants would leave their catches at the docks or with the judging booths because they did not particularly like eating fish. Fishing regulations prevented the sale of fishes caught in the tournament, and dead fishes could not be returned to the water, so aside from eating them, there was not much else that could be done with them. We gladly claimed many of those for feeding the scientific crew, and we ate a wide variety of fresh fish every day. I tasted many species, including some that you don't normally think of as food. One of these, triggerfish, became my favorite out of the forty or so edible species we had access to. While I was there, we also ate wahoo, cobia, tuna, mackerel, flounder, snapper, bluefish, marlin, grouper, pompano, shark, and stingray. I was particularly fascinated by the beautiful triggerfish because many years ago I had bought a small one as an ornamental fish for a marine aquarium, and it had cost more than $100.

I was fortunate to be at the rodeo on the day a 385-pound Warsaw grouper was brought in (see figure on page 104). This was the second largest specimen of this species ever recorded, and the silver-tongued Willy managed to talk the owner out of the skeleton in return for our expert filleting job. Grouper fillets were selling for over $10 per pound at the time, so to the owner the nearly 200 pounds of fillets was food for personal consumption worth thousands of dollars. To us, the skeleton was priceless. Very large adult Warsaw groupers were extremely rare in museum collections. It was a new experience for me to dissect fishes with sharpened machetes and hatchets rather than scalpels and fine scissors.

Dissecting the giant grouper revealed an epic story about its life and eventual capture from the deep sea. We found a dozen older fishing lures in its stomach, indicating that it had broken the fishing lines and gotten away from potential captors at least twelve times in the past! The fishermen who hooked it this time had been hopelessly towed around the ocean for hours in their small boat before they were pulled near a much larger boat with a winch. Seeing an opportunity, the fishermen asked the larger boat for assistance and passed them the line with the fish on it. The crew of the larger boat put the line on their power winch. It was only with the help of this powerful mechanical device and the larger fishing vessel that they were able to haul the fish onto a boat and ultimately to shore, where a crane could lift it out of the boat. If these fishermen had not crossed paths with the large boat, they would have eventually been forced to cut the fishing line to save their small boat from being towed around forever. The Warsaw grouper is currently listed as endangered by the American Fisheries Society, but once one is brought up to the surface from a great depth, there is no lifesaving catch-and-release possible. Large adults of this species live as deep as 2,000 feet below the surface. The mere act of bringing them to the surface usually kills them due to the extreme pressure change. At least this behemoth would be immortalized for science.

The top prize at the rodeo that year was $30,000, and there were also many other lucrative awards for the various categories of winners. As in all things that involve monetary prizes, there is sometimes a bit of dubious ethical behavior. Willy had observed several instances of this over his years at the rodeo. There had been instances of stuffing fishes with lead in order to increase their weight, which in at least one case had embarrassingly fallen out onto the scale while the fish was being weighed. One time someone entered a fish that had been caught in Costa Rica and tried to claim that it had been caught in the contest area. Another time in an effort to earn some extra funding for his church, a clergyman tried to enter an amberjack he had stuffed with many frozen bait fish to increase its weight. The cheating eventually got bad enough that the tournament introduced a polygraph test for entrants who were contenders for a prize, and an electric freshness meter for the fish (yes, there is such a thing). The first year that the polygraph was introduced, a potential winner failed the test.

After the three days of the tournament, Willy was utterly exhausted. I was largely an observer over the course of the event and experienced only a taste of what is involved in his operation, but even I was tired. In addition to begging contestants for their prize catches for a solid seventy-two hours, Willy had managed a large crew of highly enthusiastic students, volunteers, and visiting ichthyologists in dissecting much of the flesh off hundreds of fish skeletons to be shipped to UMass and the Field Museum for final cleaning by flesh-eating dermestid beetles.[2] The giant fish-gut trench was filled and buried, partially de-fleshed skeletons were boxed for shipment, and accidental cuts and punctures on the crew's hands from filleting knives and fish spines were washed and bandaged. It was time to go. It had been a particularly successful trip with hundreds of fish skeletons collected for work on rayfin fishes, including the skeleton from the 385-pound grouper.

Together, Willy and I made new scientific discoveries, trained graduate and postdoctoral students, and produced numerous technical books and scientific journal articles as a result of our collaborative research program. Our work was reviewed in one of the world's top academic journals *Science* as "a model of what should be done to carry comparative biology to the level of technical and methodological perfection that will allow the discipline to proudly enter the 21st century."[3] I refined my drawing skills during this period to illustrate anatomical details of the skeletons we worked on in publications. Anatomy was still something I found to be very aesthetic, and as both Colin Patterson and Donn Rosen told me while I was a graduate student, in order to really see anatomy, you must draw it. I traced photographs of skeletons with black ink on velum and then photostatically reduced the tracing by 50 percent or more. This produced very detailed-looking line drawings. I came to enjoy publishing combination plates that included those drawings paired with photographs of the same exact size in my monographs and journal articles on fish anatomy (e.g., page 109). It was a process that was both intellectually fulfilling and aesthetically satisfying to me. My work with Willy added greatly to my scientific reputation and was one of the most enjoyable periods of my professional career.[4]

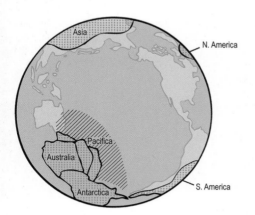
225 million years before present

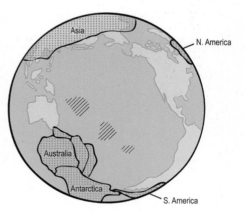
180 million years before present

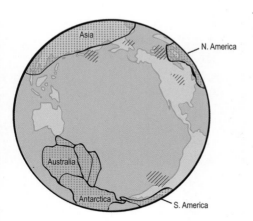
135 million years before present

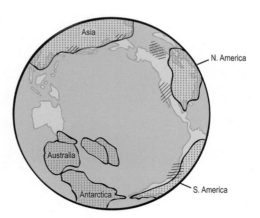
65 million years before present

The Pacific Hemisphere at four points in the distant past showing the slow breakup, movement, and coastal collisions of the lost continent of Pacifica. Past positions of Pacifica shown with diagonal lines, past positions of other continents shown in stipple, and relative positions of present-day continents shown in green for geographic context. Based on Nur and Ben-Avraham (1977).

Eugenia Sytchevskaya and me in 1994, standing in front of the Paleontological Institute of the Russian Academy of Sciences (PIN), Moscow. Eugenia was the curator of fossil fishes for the PIN and helped Willy and me see fossils that no one outside of Russia had ever viewed before.

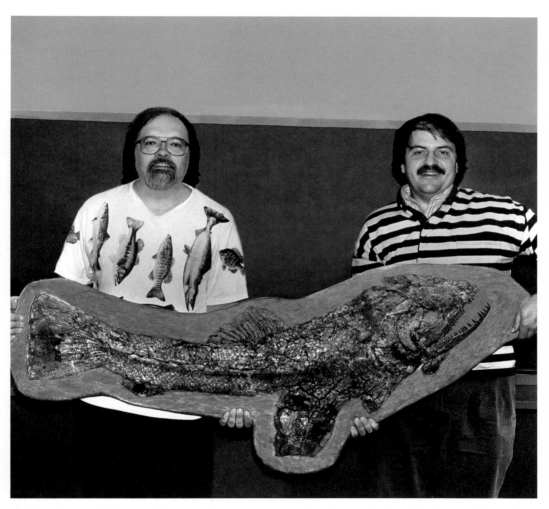

Willy and me in 1995 at the Kitakyushu Museum, Japan, studying their collection of fossil rayfin fishes. This particular piece was a bowfin from 110-million-year-old limestone in eastern Brazil. Our study showed its closest relative to be from western Africa, indicating the different position of continents during that geologic time period.

Willy and the 385-pound Warsaw grouper (*Epinephelus nigritus*) at the 2002 Alabama Deep Sea Fishing Rodeo. It was the second largest specimen of the species ever recorded at the time of its capture, and we were given the skeleton for science in exchange for filleting the fish for the owner.

Willy and me dissecting a 300-pound marlin at the 2002 Alabama Deep Sea Fishing Rodeo. When dealing with large "trophy-sized" marine fishes, dissection requires machetes and hatchets rather than fine scissors and scalpels. It is a job not for the weak or faint-hearted.

Student volunteer at the Dauphin Island Sea Lab who has dissected most of the flesh off of a barracuda skeleton before it is packed for shipment to the Field Museum in Chicago. Once back at the museum, this and other partly de-fleshed fish skeletons were placed in tanks with flesh-eating dermestid beetle colonies for final cleaning.

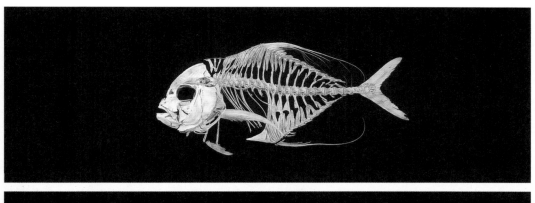

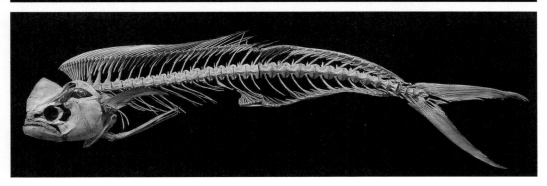

After the final cleaning by the beetles, the barracuda from the last page (*middle*) and other rayfin fish skeletons such as this pompano (*top*) and this mahi-mahi (*bottom*) were rinsed, dried, and made ready for study and incorporation into the museum's permanent collection. We used this method to make many large skeletal preparations from specimens collected at the fish rodeo.

Willy and I also collaborated on training doctoral students. One of our students, Eric Hilton, is now curator of fishes for the Virginia Institute of Marine Science (VIMS) and is regarded as one of today's leading fish anatomists. Here he is holding a specimen of the American shad, one of over 500,000 alcohol-preserved fishes in the VIMS collection.

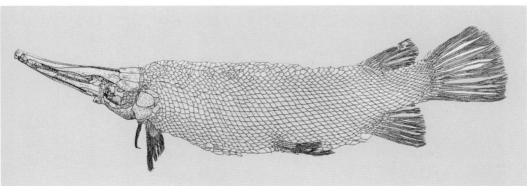

Photograph and line drawing of *Lepisosteus bemisi*, a garfish species that I described and named after Willy in 2010 in remembrance of our many collaborative exploits. This beautifully preserved specimen is two and a half feet long and is from my 52-million-year-old field site in southwestern Wyoming (chapter 3). This was one of hundreds of combination figures that I published over the years describing fossil and living fish skeletons.

The iconic 42-foot-long skeleton of SUE, mounted in the main hall of the Field Museum. This specimen ignited many emotionally and politically charged issues. My involvement with SUE added another important layer of experiences to my development as a curator.

6

A Dino Named SUE

It was more than mentors, colleagues, and fieldwork that shaped my curatorial career—it was also extraordinary events. One of the most memorable trials of my curatorial career, both figuratively and literally, involved the museum's most prominent icon: SUE the dinosaur.

Tyrannosaurus rex is the most popular of all dinosaurs, and the specimen called SUE is the largest, most complete *T. rex* skeleton in the world. This fossil has become the icon of Chicago's Field Museum, and it stands as a regal centerpiece in the museum's main entry hall. Officially, it is specimen number FMNH PR2081, only one of more than 27 million pieces in the museum's collection. But what a specimen! Even its name is trademarked (in the form of having all capital letters). Since arriving in Chicago, it has been a never-ending source of scientific discovery, public outreach, and museum promotion. The year that SUE debuted at the museum, the attendance jumped to 2.4 million, an increase in 900,000 visitors over the previous year.

The history of SUE's discovery, excavation, federal seizure, and eventual sale to the Field Museum is one of the most extraordinary paleontological sagas of all time. It involved curators from all over the country as well as a number of other colorful individuals. It changed the lives and fortunes of many people. This epic story has been the subject of at least two major books, several shorter-length children's books, a NOVA special called *Curse of T. rex*, and an Emmy award–winning movie, *Dinosaur 13*. The story has been portrayed from many perspectives by commercial fossil dealers, attorneys in law journals, TV reporters, video documentarians, federal agents, newspaper journalists, and politicians. As someone who was involved early on in the episode and whose home institution ended up acquiring SUE for its permanent collection, I can provide yet another inside perspective on the story: that of a museum curator caught up in the frenzy.

A quick sketch of SUE's early history would go something like this: In life SUE stood over 13 feet high at the hip, 42 feet in length, and probably weighed close to 10 tons. It had the most powerful bite force of any known land animal, fed on other dinosaurs including smaller members of its own species, and could eat hundreds of pounds of flesh in a single meal. This predatory titan survived several major injuries thought to have been caused by struggling prey (broken ribs, torn tendons, and a damaged shoulder blade) and a variety of ailments (arthritis, gout, and infections) until it died at the ripe old age (for a *T. rex*) of twenty-eight. I say *it* rather than *she* because even though the specimen's nickname is SUE, its sex has never been conclusively determined. Fortunately SUE died close enough to water to be buried with silt and sand in a relatively short amount of time. This protected the body from scavengers and enabled the bones to become mineralized and well preserved over epochs of geologic time.

Although SUE appears to have had a tough life, you could say that its real troubles began after its discovery as a fossil some 67 million years later. Less than two years after being found and excavated, it ignited the most extensive legal battle of any fossil in history. It turned out to be the catalyst for the largest criminal case ever tried in South Dakota. The investigation lasted over five years, included two grand juries, and ended with a jury trial lasting eight weeks. There was a bitter dispute between the U.S. federal government and a commercial fossil company named the Black Hills Institute of Geological Research (BHI), which spilled over into several major museums and involved museum curators from around the country, including me. The trial ended up nearly bankrupting the BHI in legal fees and sending its founder and president, Peter Larson, to federal prison on charges that many South Dakotans came to regard as an abuse of prosecutorial power. It is a case that has many parts and is used today by law schools around the country in classes on cultural property statute. To establish the context of the dispute, it is necessary to first to explain what the BHI is and how it differs from an academically based, nonprofit natural history museum.

The Black Hills Institute of Geological Research, Inc., was founded in 1974 by Peter Larson and is located in Hill City (population about 700 in the year SUE was discovered). The BHI is a business that acquires and sells fossils, minerals, and casts of fossils. It also maintains a small museum with a separate exhibit and research collection that is not intended for sale. The organization has stated that they eventually hope to spin off the museum into a separately incorporated nonprofit entity called the Black Hills Museum of Natural History, but so far this plan has not come to fruition. The operating expenses of the BHI museum are currently funded by its commercial operation. The fact that the BHI's primary business is selling fossils but it calls itself a research institute has often made it a touchstone for controversy within the world of professional paleontologists. Some see the BHI as a commercial operation that is misrepresenting itself as a legitimate academic institution. Others see the BHI as a capitalistic start-up venture that has been unfairly attacked by intellectual elitists before it has had the chance to get off the ground. There are extreme views on both sides of the issue that still simmer today. The recent movie *Dinosaur 13* exacerbated the conflict further. At times this movie gratuitously attempts to cast academics as existing within fiefdoms and lacking the field expertise possessed by commercial fossil hunters—a point that after almost forty years of my own fossil expeditions, I take serious issue with. Museum curators have been collecting dinosaurs in areas ranging from the deserts of Mongolia to the mountains of Antarctica for decades. They work with federal agencies to construct collecting policies for public lands to ensure that collecting there is well planned and professionally conducted. The goal for professional (academic) paleontologists is to insure that important specimens are placed in public trust so that their story can be studied now and in the future. *Dinosaur 13* is no documentary. It is an exploitative attempt to show unbroken admiration of the BHI as the David vs. Goliath underdog at the expense of everyone else. Too bad. The embellishments in the movie were totally unnecessary because the real drama of the SUE confiscation and the distorted turns in the case against the BHI could have made good cinema without exaggeration or journalistic bias.

I first met Peter Larson in the late 1970s when I was working on my master's at the University of Minnesota. He had sent me pictures of some important fossils in the BHI museum collection, and I wanted to use images of them in the book I was preparing on the paleontology of the Green River Formation. He stopped in Minneapolis to personally drop off the specimens. I was grateful that he personally delivered the specimens for me to study, and I was impressed that someone in South Dakota would attempt to build a museum from scratch in this day and age. Although our core objectives came from different angles (mine academic, his commercial), he had a likable and generous personality, and over the years we occasionally corresponded about various specimens in the permanent BHI collection. He sported a bushy mustache, glasses, and the sort of weathered tan skin that people in the western United States often have. Even today in his early sixties, he somehow maintains a country-boyish, everyman look that endears him to the media and helped make him a local hero in South Dakota. To some degree he also comes across as a regular paleontologist, but there is no question that the business of the BHI is his priority.

The fossil collection of the BHI museum contains some uniquely important fossils that are made available to the scientific community for research. Over the years I have borrowed several of these specimens for study, and on one occasion I borrowed several fossils from the BHI for a temporary exhibition I helped design and facilitate at the Field Museum called *Locked in Stone: The Prehistoric Creatures of Fossil Lake*. I also purchased a few important pieces for the Field Museum's collection from the commercial arm of the BHI. They routinely sold many specimens to natural history museums around the world, ranging from small university museums such as Harvard's Museum of Comparative Zoology to the Smithsonian Institution's National Museum of Natural History. Although it is disingenuous to call the BHI a research institution, Larson was always cooperative and helpful to the scientific institutions with which I have been associated. But my early relationship with Larson and the BHI for the Field Museum transactions ultimately drew me into a challenging ordeal. I was pulled into what began as the SUE trials, as were other curators and professional scientists from around the country.

SUE the dinosaur was so named by Larson after the woman who discovered it, Susan Hendrickson. Hendrickson has been described by various museum curators around the country as "field paleontologist," "marine archaeologist," "explorer," "intrepid globe-trotter," "treasure hunter," and a host of other names ranging from critical snipes to unabashed admiration. She never graduated from high school but passed the GED high school equivalency test. She never went to college but has an honorary PhD from the University of Illinois at Chicago. She is not a scientist, but she is an accomplished explorer. Between 1998 and 2001, she appeared several times at the Field Museum as a celebrity guest to talk about her discovery of SUE to the press. Hendrickson has an innate talent for finding things of importance. What she lacks in professional training she makes up for in focused persistence, good fortune, and charm. In the words of dinosaur curator Bob Bakker[1] from the Houston Museum of Natural Science, "There are some people who can find fossils and some who can't. Susan has the talent for getting a sense of place in a paleontological context. You must be born with it. She has it." On August 12, 1990, Hendrickson made the paleontological discovery of a lifetime: the world's most famous dinosaur specimen.

Hendrickson loved to do fieldwork. In a sense she *lived* to do fieldwork. When she wasn't looking for dinosaurs with the BHI, she was doing underwater salvage work off the Florida Keys, searching for amber fossils in the caves of the Dominican Republic, or working with a team of marine archaeologists in the Middle East. She also visited my field site in Wyoming during the early 1980s, which is where I first met her. You might say she had the passion for exploration and fieldwork that curators have, but she herself would admit to lacking the passion for research and analysis that curators must also have. On the morning of her greatest discovery, she had been on a long prospecting expedition for dinosaurs with the BHI on a couple of private ranches, one of which belonged to a rancher named Maurice Williams. It had started out as a somber day. She recounts in Steve Fiffer's book *Tyrannosaurus Sue* that she and Peter Larson had decided to end their romantic relationship with each other some days earlier. She opted to stay behind by herself to continue prospecting while Larson and the rest of the crew went to the nearby town of Faith, South Dakota, to fix a flat tire. In the distance about seven miles away from camp, there was a sandstone cliff that had caught her eye a couple of weeks earlier. She set off on foot, and by the time she reached it, several hours had passed. She stood before the sixty-foot buff-colored hill and began looking for bits of bone. As she scouted along the ground at the base of the cliff, she came to some small bone fragments that had rolled down from higher up. As she looked up the hillside to find where they had come from, she spotted three large vertebrae and a leg bone beginning to weather out of the cliff face. She knew immediately from the appearance and position of the large bones that there must be more of the skeleton still in the hill. She became even more excited as she noticed that the bones she found lying near the base of the hill had a very distinct appearance, indicating this was a carnivorous dinosaur. She collected some of the bone fragments and ran almost the full distance back to camp to tell Larson, who by then had returned from Faith. After showing him the small bone fragments she had collected, he recognized them as being from a T. rex. They hurried back to the cliff face, running much of the way. Once they arrived, Hendrickson reportedly pointed to the skeleton and said, "This is my going-away gift to you." Larson told Hendrickson then and there that he would name the skeleton after her. It was a joyous moment in what would gradually develop into a tragically convoluted saga for both of them.

For the next several weeks, Larson, Hendrickson, and several other members of the BHI labored to excavate the skeleton of SUE the T. rex. The bones were buried under almost 30 feet of siltstone, sandstone, and sand that had to be removed by hand (no tractors) in order to retrieve the blocks containing the skeleton. Larson and three others did the excavating. They had to hike in each day because Maurice Williams would not permit them to drive over his land. It was difficult work that had to be done with picks, shovels, and knives in blistering heat, with daily temperatures exceeding 100°F. They worked long twelve-hour days every day for more than two weeks to expose the blocks of siltstone containing the bones of SUE. After most of the bones were superficially exposed, Larson gave Williams a check for $5,000 for the dinosaur in the ground (a transaction whose intent was later disputed by Williams). Then the careful excavation process continued. Each bone was mapped in place for documentation before it was removed. By all reports (e.g., Bob Bakker, and what we at the Field Museum later determined), the excavation job was of professional caliber. The BHI made a detailed site map of all bone positions in the ground before they were excavated. Williams and his family members would sometimes come to watch the progress of the excavation. Once the overlying rock layers were taken down and the bone-filled blocks were exposed, they were covered with plaster jackets supported with two-by-four wood framing for safe transportation back to the BHI facility. The jacketed blocks were huge, and the largest weighed nearly 10,000 pounds. On the last day, the BHI crew was given permission to drive vehicles over Williams's property to transport the fossil-containing blocks from the ranch back to the BHI fossil preparation facilities in Hill City, South Dakota. The crew loaded the blocks onto several vehicles, including a five-ton trailer with the aid of a winch. Then, on the last day of August 1990, the unprepared skeleton of SUE the *T. rex* left for Hill City.

Once the specimen was unloaded at the BHI, preparation of it began immediately. One of the BHI fossil preparators, Terry Wentz, was assigned full-time to SUE's preparation. Larson, his brother Neal, and Robert Farrar also put in time on its preparation when they weren't tending to the business side of things. Over the next twenty months, dozens of stories about the huge *T. rex* appeared in newspapers and magazines. Within the rock being removed around SUE's bones, other fossils were discovered: including plants, turtles, and a bone from a baby *T. rex*. Fossils were sent to curators and other researchers around the world for identification and analysis. At the October 1991 meeting of the Society of Vertebrate Paleontology, Larson gave a presentation on what had been uncovered from the excavated SUE material so far and invited paleontologists to come to Hill City and study the material. By early 1992, thirty-four paleontologists had signed up to collaborate on a *T. rex* monograph featuring SUE. The skull and upper arm were among the most important parts that had been prepared. SUE was only the second *T. rex* known with the upper arms preserved, and the upper arm of SUE was the best preserved of the two by far. It was loaned for study to curator Ken Carpenter from the Denver Museum of Nature and Science, who, based on the SUE arm, demonstrated that the forearms of *T. rex* were twice as powerful as had had previously been thought. Although proportionately small, they were stout and muscular, and those of SUE could have lifted hundreds of pounds at a time.

But not all was harmonious among the paleontologists and other interested parties regarding SUE, and there was suspicion about the motives of the BHI. The fact that the BHI sold fossils commercially put them in conflict with many in the professional scientific community who saw the private ownership of scientifically important fossils as unethical. Also, there had been previous allegations that the BHI had collected fossils illegally from federal lands. The Society of Vertebrate Paleontology (SVP)[2] was pushing for tighter federal regulations for fossil collecting on federal lands, while the BHI was lobbying hard for looser restrictions.

The SVP considered vertebrate fossils to be a significant nonrenewable resource that when found on public land should be held in the public trust, preferably in a museum or research institution, where they can benefit the scientific community as a whole. The BHI's stance was that if fossils in the South Dakota Badlands and elsewhere are not collected once exposed to weathering, they are of no use to anybody because they are rapidly destroyed by that weathering. They argued that there were too few professional paleontologists to collect and preserve fossils currently exposed to the elements. They supported a 1987 National Academy of Sciences (NAS) Committee recommendation to the U.S. Secretary of the Interior that agencies allow commercial collecting through a regulated permit system. (Larson was a member of that committee.) This report had been done at the request of the U.S. Department of the Interior, but the Secretary of the Interior rejected the NAS recommendation.

The controversy continued to heat up. The SVP representation of the day (the executive committee) was largely critical of the BHI, although there had been divisions of opinion within that group culminating in the resignation of its vice president and Smithsonian curator Clayton Ray. Professional paleontologists were lining up on both sides of the issue. Bob Hunt, secretary of the SVP and member of the SVP Government Liaison Committee, was quoted as saying: "Officers of the Black Hills Institute have not published scientific studies demonstrating serious scientific expertise in dinosaur research, nor are they able to identify the best experts to undertake such a study." In response, Harvard-educated dinosaur paleontologist Bob Bakker backed Larson as a responsible paleontologist and referred to the criticism of the BHI as a "campaign of slander."

By 1991 the troubles for the BHI and its senior staff were beginning to mount in both criminal and civil court. Federal officials had been receiving multiple reports of illegal fossil collecting on federal lands and were building a criminal case that would include charges that SUE was also illegally collected. Maurice Williams also decided that he had been "taken" when he agreed to sell the bones for $5,000, even though he had cashed the check and was on videotape acknowledging the transaction. He decided to challenge the deal with the BHI in civil court, claiming that the payment only entitled the BHI to remove and prepare the fossil for later sale. In addition, the Sioux Nation had also become involved because Williams was a member of the Sioux tribe, and his ranch was within the Cheyenne River Sioux Indian Reservation. They claimed that Williams did not own the fossil but that the entire tribe did. Steve Emery, the Sioux lawyer representing the tribe, was quoted in a newspaper article saying, "Our reservation has the 14th and 29th poorest counties in the country. We do not have a tribal casino here. We thought the dinosaur fossil could have been the foundation for a [tourist] museum." Williams would have to engage his own tribe in a legal battle over the dinosaur. There were now three different entities claiming title to SUE: Maurice Williams, the BHI, and the Sioux Nation.

Back in the little town of Hill City, the preparation of SUE's skull was progressing rapidly after twenty months of work. The surrounding community was enthralled by the process. Over two thousand visitors had come to see the skull being prepared in the back rooms of the BHI museum. The number of spectators was more than twice the entire population of Hill City itself. Because of the increase in tourism in the town as a result of SUE, many local residents started viewing the fossil as a piece of local heritage. Scientists and journalists from around North America continued to stop by to see it. As the preparation continued on the articulated skull, the BHI decided to use some new technology to look at the internal anatomy of it: CT scanning. The skull was scheduled to be shipped to NASA's Marshall Space Flight Center in Huntsville Alabama in May 1992 for scanning. Before that could happen, the situation exploded.

In a surprise raid at seven o'clock on the morning of May 14, 1992, a group of thirty-five federal agents plus thirty National Guard soldiers descended on the Black Hills Institute to seize SUE in its entirety along with other fossils and many boxes of documents and correspondence. A fourth entity was now claiming ownership of SUE: the federal government. Initially, the feds believed that they might own the skeleton because the U.S. Secretary of the Interior held Williams's land in trust at the time the dinosaur was excavated. It took three days to pack up and cart away all the fossils and paperwork. As the FBI, National Park Service, and National Guard moved the material into large military trucks, a large crowd of local residents started to grow. By the third day, there were nearly two hundred protesters (out of a town of only seven hundred) chanting phrases like "Shame on you" and "Don't be cruel, save SUE." Emotions were high, and even the mayor of Hill City became involved. It was an inflammatory state of affairs for a politically conservative population that was instinctively suspicious about federal government intervention in local matters. The raid for SUE had been organized by acting U.S. attorney Kevin Schieffer, who had only been on the job for a week before he was asked to look into the case. It also keyed into earlier investigations by the National Park Service into the activities of the BHI.

Before SUE came into the picture, the federal government had already been looking at the BHI. Beginning in 1985, the National Park Service had begun investigating allegations and evidence that the BHI had been linked to unauthorized fossil collecting on federal and Indian lands. This investigation was led by National Park Service ranger/paleontologist Vince Santucci, who had been investigating the BHI for six years before the skeleton of the *T. rex* named SUE was discovered. Santucci was the chief ranger at Fossil Butte National Monument, and I had met him on several occasions when I was doing fieldwork in Fossil Basin, Wyoming. He had strong convictions about protecting paleontological sites in national parks from poachers. During his work for the National Park Service in the mid-1980s, he had become concerned about the widespread theft of fossils from national parks, including Petrified Forest National Park in Arizona and, most notably, Badlands National Park in South Dakota. He was quoted by journalists, comparing fossil thieves to grave robbers and looters, and was dubbed by the superintendent of Petrified Forest National Park and members of the press as the government's only "Pistol-Packing Paleontologist." In his investigation of fossil theft from the Badlands, Santucci kept coming across the name of the BHI as well as a few specific individuals. The National Park Service had been seeking assistance from the U.S. Attorney's office for several years without success concerning fossil theft activities in Badlands National Park. But once the U.S. Attorney's Office in Pierre, South Dakota, began the SUE case, the National Park Service's investigation at Badlands National Park was rolled into it, much to Santucci's dismay. The criminal and the civil trials concerning SUE and other fossils were both conducted with the same attorneys using the same evidence collected during the SUE seizure. This led to confusion by the public over the individuality of the cases, and the National Park Service case became completely overshadowed by the media and public focus on the dinosaur.

Schieffer thought that using SUE as the focal point of the federal case would insure that it received wide attention, which it certainly did. It is reported that he may have also had personal motives beyond concern for fossil thefts involving his political ambitions. But the shock-and-awe strategy of the federal authorities ultimately backfired in the conservative state of South Dakota. The local mood in Hill City and eventually Rapid City was that this was a case of government heavy-handedness toward one of South Dakota's own. As the federal trucks pulled away with the bones of SUE, parades of protesting Hill City residents continued to voice their concern, and one person was even caught on video lying down in front of one of the federal trucks in an effort to stop it.

The investigation that followed the raid included the U.S. Department of Justice, the Federal Bureau of Investigation, the Internal Revenue Service, the U.S. Customs Service, the Bureau of Land Management, the National Park Service, the U.S. Forest Service, and the Bureau of Indian Affairs. It seemed an amazing conglomeration of federal agencies and a serious investment of U.S. tax dollars. A year and a half later in November 1993, after many months of intensive effort led by the U.S. Attorney's office, the BHI and several individuals within the organization were served with an incredible indictment specifying 39 counts and a potential 153 convictions.[3] For Peter Larson, founder of the BHI, the charges added up to more than 350 years in prison and over $13 million in fines. The feds were not messing around! And again, the mood of many South Dakota residents was one of agitation. Larson claimed, through several media outlets, that the government was seeking more prison time for him than had been received by serial killer Jeffrey Dahmer for killing and eating fifteen people!

At the Field, we first became aware that museums were being pulled into the case in early March 1993. The FBI had previously served a subpoena to one of the Smithsonian's curators of fossil mammals, Clayton Ray, to appear before the grand jury in Rapid City to answer some questions concerning the BHI. Ray had testified and government prosecutors had reportedly been very aggressive with him. This had come as a surprise to the Smithsonian because the expectation had been that since they were the National Museum of Natural History and this was a federal case, there would be a cooperative respect for Ray's voluntary testimony. Word of this turn got back to the museum community, as did news that there were subpoenas shooting all over the country and that we could expect one at the Field too. Among the documents the government had confiscated from the BHI were copies of all their correspondence over the years, including letters between Larson and me going back to my time at the University of Minnesota. The federal attorneys took note of all those who had any dealings with the BHI to put together a potential list of witnesses for the prosecution. Schieffer was playing hardball, and he seemed to feel that anyone or any place that had dealt with the BHI was hiding information that could be used in his case. Based on the indictment against the BHI and the scale of the government's case, Schieffer appeared to be betting the farm that he would nail them.

On March 15, an FBI agent came to the Field Museum to serve three subpoenas. One went to museum president Willard "Sandy"

Boyd for documents. The other two went to me and the museum's archivist for an appearance before the grand jury in Rapid City. We met in Sandy's office. In attendance were the museum's general counsel Richard Koontz, Sandy, me, and the FBI agent serving the subpoenas. The agent was extremely polite. For me, who had never seen anything quite like this, it made quite an impression. He asked several questions that I answered.

"Have you ever heard of Black Hills Institute?"

"Yes I have."

"What sort of contacts did you have with them?"

I borrowed specimens from them for research on occasion and purchased a few specimens for the museum. I visited their shop once in Hill City."

"Did you ever do any fieldwork with them?"

"No."

"Do you have any knowledge of catfish fossils from South Dakota?"

Yes, I purchased some for the museum. They represent a new species and will probably represent a type series for a scientific description someday."

"Did you buy fossils from Black Hills Institute for the museum or for personal ownership?"

"No, only for the museum."

"What is a type series?"

"The specimens on which a newly named species is based."

"How long have you known Peter Larson and the other people at Black Hills Institute?"

"About ten years or so."

"Where can I find Leigh Van Valen and David Raup [to deliver additional subpoenas]?"

"Not sure. Leigh Van Valen and David Raup are both professors at the University of Chicago. Why don't you check with the university?"

And then it was over. Or I should say it was just the beginning for me and the museum's archivist. The meeting with the FBI agent had started a long process that would take me to South Dakota on three separate occasions. Sandy Boyd is an intelligent, cautious man who had an experienced knowledge of the law. He had been dean of the Law School at the University of Iowa and then the university's president during the years 1969 to 1981 before coming to the Field Museum. He was also very protective of the museum and its employees. Sandy retained a top-notch attorney, Nathan P. Eimer (Nate), from the firm of Sidley & Austin in Chicago. Sandy said that Nate would accompany me and the archivist whenever we went to South Dakota, and he would represent the museum, the archivist, and me. Nate turned out to be invaluable during the amazing events that were unfolding.

My first trip to Rapid City to testify before the grand jury was in early October 1993, where we were planning to be for three days. We checked into the Ramada Inn on Monday, October 4, to be ready for the meeting the next morning. Now there is not a lot to do in Rapid City, but we were determined to make the most of it to relieve the building tension. Nate and I explored downtown, and in a few hours we pretty much saw all there was to see. Toward late afternoon I remembered that we were next to one of the country's most spectacular fossil sites, Badlands National Park. Given that it was fossils from the Badlands that had helped generate the investigation of the BHI to begin with, it seemed appropriate that we might kill some time seeing the park. It was famous for containing the richest deposits of Oligocene mammal fossils of anywhere on Earth. Outcrops of the Brule Formation provide a unique look at life in the region some 30–33 million years ago, and there are fossils weathering out everywhere. In the developed area of the park, there are fossils that have weathered out and are now covered with clear plastic domes protecting them from the elements (and souvenir collectors). These exhibit capsules are situated along convenient pathways so people can get a close look at them. While we walked through the Badlands, Nate asked me questions about paleontology and I asked him questions about the federal grand jury system. Nate instructed me on what was going to happen during the proceedings. The grand jury is a legal body empowered to investigate potential criminal conduct in order to determine whether criminal charges should be brought and whether the target (the BHI in this case) should be tried. The grand jury system is a vestige of the English legal system dating back to an 1166 act of King Henry the Second. It is not a court, and subjects of the investigation (such as museum curators Clayton Ray, myself, and others) would not be allowed to have attorneys present in the grand jury room. Nate said he would have to wait outside until I was finished. Once I came out afterward, I would relay to him what had happened.

Part of what Nate insisted on was that the museum archivist, the museum, and I get an order of immunity from the U.S. District judge before we would agree to testify or before we would say anything at all to the federal prosecutor. He had some considerable experience with the federal court system, and he explained that a common tactic of an aggressive prosecutor, such as Kevin Schieffer, would be to invite a subject into the grand jury room and then ask them a series of questions for which an assistant at the back of the room would write down the answers that were given. He would then ask the subject to return the next day and ask pretty much the same questions. If any of the answers given on the second day deviated from those of the previous day, the federal prosecutor would then implicitly threaten the subject with a charge of perjury unless the subject gave additional information against the case target (the BHI in this case). Nate relayed that this generally happens when a prosecutor believes that the subjects are withholding information about a target. The trouble was, I had no additional information about the BHI. Nate warned that until the order of immunity came through, the archivist and I should answer no questions, strike up no conversations, or even give Schieffer the time of day. "Just read the words on this card," he said. He then wrote out the words, something to the effect of "on advice of counsel, I refuse to testify in accordance with my rights under the Fifth Amendment of the United States Constitution." This all seemed pretty melodramatic to me. Why should I not fully cooperate with the federal government? They are the good guys, right?

And it was difficult for me to believe that something like this could happen right here in Rapid City. But Nate was persistent and spoke in a way that conveyed knowledge born of experience.

The next morning Schieffer invited me and the museum's archivist behind closed doors. I took a seat on the other side of the desk from the energized prosecutor. The archivist sat to my right. Then Schieffer asked me how my hotel was and was this the first time I had been to Rapid City? I turned around and looked behind me to see another man writing notes. Nate's words from the previous day describing such a scene then struck me, and I felt my face turn red. I pulled out the card and read it. This time, the federal attorney's face got red. He asked me another question, and I gave the same response. He finally looked at me and the archivist and said something to the effect of "You should get a comfortable place here in Rapid City because you will not be excused from Rapid City until I see you a second time and that meeting might not happen for a while." Now we were getting concerned. I did not want to be spending the next month or two sequestered in Rapid City, South Dakota. The archivist looked visibly shaken. When we left, we found Nate waiting in the hallway and told him what had happened. He told us not to worry. It was just an intimidating attempt to get information that the attorney thought we were hiding. Nate said his firm would call a senator from South Dakota the next day and we would be out of there in no time. Nate was right. The next morning we were given our order of immunity (which had actually been filed more than a month earlier but not revealed to us when we first arrived). We hadn't even needed a second meeting with Schieffer, and we were given permission to leave Rapid City. At this point I came to appreciate the fact that the archivist and I had good legal counsel with political connections, and I wondered what happened to folks who did not have the resources to come with an attorney. We went back to Chicago, and I would later come back to Rapid City two more times for the case.

Months later when I returned to Rapid City to go through witness preparation for the trial, Kevin Scheiffer had been removed from office. Technically, Schieffer had been only the acting U.S. Attorney when he ordered the raid on the BHI, and his appointment to U.S. Attorney had later not been approved by the Clinton administration. The case had been blown way out of proportion. Schieffer had been replaced by Robert Mandel, who appeared to want to get the government out of this case in as face-saving a way as possible, while still looking for any legitimate violations of the law. I could only imagine what sort of damage control was being sought at this point for a case that had become sort of a PR nightmare for the government. Mandel asked me one more time if I had anything else to say about the BHI that would be applicable to the case, particularly considering that an order of immunity existed for me and the museum. I did not and told him so. He walked me through what would happen when I was called to testify at the trial, and it was pretty simple after that. He was quite friendly, and he happened to mention that the government would be returning the skeleton of SUE back to Maurice Williams and encouraging him to sell it. The Sioux Nation had dropped their claim to the dinosaur, saying they now considered it the property of Maurice Williams. At that moment, some unexplainable premonition made me think that SUE was destined to end up in Chicago. I kept my excitement in check, but the hopeful possibility embedded itself in my mind.

Overall, the case against the BHI was not going well for the government, and it seemed that the prosecution was grasping for anything that might stick in order to justify the enormous amount of taxpayers' money that had been wasted on this case. What had started in the media as the "SUE trials" had now turned into something else completely. I went back to Rapid City one more time for the trial itself, and I spent about a minute on the stand as one of ninety witnesses for the prosecution. They asked me for incriminating pieces of information against the BHI that they appeared to think I might still somehow remember. But there was really nothing more I could say, and as I spoke I wondered why they even had me show up. And that was the last of my participation in the federal case against the BHI. I returned to Chicago the next day.

The trial continued until March 14, 1995. Then, after nearly three weeks of deliberation, the jury voted not guilty on almost all 39 counts and 153 possible convictions. In the end, the only felony that Peter Larson was personally convicted of was transporting money in and out of the country in excess of $10,000 without filling out the proper customs form. Most significantly for the government (and Vince Santucci, in particular), there was one felony conviction against the BHI for theft of fossils from Badlands National Park.[4] Even in its diminished scale, this was a landmark case of sorts. It was the first felony conviction for fossil theft in U.S. history through a criminal trial. The BHI had admitted guilt in the charges of collecting fossils illegally from federal lands, although they attributed these instances to mistakes resulting from bad reading of maps in areas where property lines were confusing, being misled by landowners about where their property lines were, and buying fossils from people who claimed to have taken them from private land.

Once the verdict was in, one thing remained: sentencing. The felony theft charge against the BHI would result in a small fine, but the felony charge against Peter Larson (failure to report to American customs officials the $31,700 in traveler's checks he had brought from Japan and failure to report $15,000 in cash he had taken to Peru for field expenses) resulted in something different entirely. There are published accounts claiming that Larson's attorney Patrick Duffy had antagonized the judge by making too many public statements to the press during the case. The defense had tried to remove the judge from the case on three different occasions in a very public way, stating that he had not been impartial on the case. It was a strategic gamble by the defense that had failed, and it looked as though the judge might be taking it somewhat personally. Larson's felony conviction for the undeclared checks was predicted by his attorney to result in simple probation of zero to six months. Duffy was filmed puffing a victory cigar outside the courthouse after the case in a show of bravado. But instead, the judge gave Larson two years in the federal penitentiary in Florence, Colorado, also called the "Alcatraz of the Rockies." He was ordered to report there on February 22, 1996. Larson ended up doing eighteen months at Florence, three months at a halfway house, and three months off for good behavior. After that he served an additional twenty-four months' probation. On the first day that he reported to Florence, the official reason for his incarceration written on the prisoner manifest was "failure to fill out forms." The guard who led him through the registration procedure into the prison looked down at the list and reportedly said, "Wow. You must have really pissed somebody off." Curator of invertebrate paleontology Niles Eldredge from the American Museum of Natural History was quoted calling the chain of events "a horrific episode in American paleontology that should never be allowed to happen again." The subject of the SUE trials are still a sore point for many involved in the trials such as Clayton Ray and the Smithsonian. When I asked Ray to comment on the ordeal for this book in January 2014, he declined, saying he was enjoined to silence by the Smithsonian lawyers at the time and has never been advised otherwise. At the SUE opening in Chicago years later, Susan Hendrickson said the government action and Larson's sentencing had greatly affected her. She was quoted from the opening saying, "It shattered my faith in the United States government." She moved to an isolated Honduran island in 1997.

The BHI played a bit too fast and loose with regard to where they collected or purchased fossils. Their conviction on the charge of theft from public property (Badlands National Park) constituted a serious offense. That charge was what had started the investigation against the BHI by Vince Santucci in the first place. But bringing SUE into the case turned out to be a great mistake by the prosecution, and the federal government's action against Larson for taking the undeclared checks in and out of the country was out of proportion. It was one more thing to add to an emotionally charged story that ultimately attracted book authors and movie producers alike. After the federal trial, the BHI became much more careful with regard to where it was collecting and buying fossils. They also became much more famous. They expanded their business in selling casts (reproductions) of fossils. Later, even the Smithsonian's National Museum of Natural History would purchase a *T. rex* cast named Stan from

Larson and the BHI to feature in its hall of dinosaurs, as would over thirty other museums around the world.

Once the civil trial was over, the bones of SUE became legally the property of Maurice Williams. The federal government had concluded that the sale of SUE to the BHI was null and void because Maurice had not received the required permission from the U.S. government before selling it previously. SUE had come from Sioux reservation land that the U.S. government was holding in trust for Williams, which meant that he did not have clear title to the specimen at the time he sold it. Now the federal government was giving him complete title to the SUE skeleton and granted him permission to sell it to the highest bidder. It was a jackpot ruling for Williams, and he was already hearing estimates of $1 million or more for sale of the fossil. In 1997 he put SUE up for auction at Sotheby's in New York. It seemed a shame that such an important specimen could end up being bought by a gambling casino or as an adornment for a billionaire's private home. A short time later, the wheels started to turn in Chicago.

When I returned from the trial in Rapid City, the Field Museum had been in the process of getting a new president and CEO. John McCarter was hired by the museum in the summer of 1996 on a platform that he was going to ramp up the museum's attendance, among other things. David Redden, executive vice president of Sotheby's auction house in New York, contacted McCarter early in 1997 to see if he could use one of the Field Museum's murals in their upcoming catalog, which was going to feature an auction of SUE the dinosaur. As part of the conversation he asked McCarter, "Your museum doesn't have a T. rex, does it?" McCarter answered, "No." Then Redden said, "You ought to have one." McCarter started thinking about it. What if SUE were to come to the Field Museum? That might be just the catalyst needed to boost the annual attendance at the museum. McCarter went to the chair of the Geology Department, John Flynn, to poll the department's curators about whether or not to make a serious effort to acquire SUE. Did we think the specimen fulfilled the three necessary criteria to support the attempted acquisition (important for the collection, important for research, and important for public education)? Our vote was an enthusiastic yes! So McCarter, Flynn, and chief fossil preparator Bill Simpson went to Sotheby's warehouse on the Upper East Side of Manhattan to see the fossil. McCarter wanted Flynn to attest to the specimen's scientific importance and Simpson to assess whether the specimen could be prepared and mounted in time for unveiling sometime in 2000 as a millennial event. The three of them examined the field jackets and boxes containing the partially prepared skeleton along with all associated documents, and then McCarter asked Simpson and Flynn, "Do you want it?" They said, "Yes." And so began the concerted efforts to acquire SUE for the museum.

McCarter began looking for major donors to finance a serious bid for SUE. He tried several companies without success until approaching McDonald's. McDonald's not only committed $5 million, but they brought in the Disney Corporation with a commitment of another $5 million. Now the museum had the backing to make a strong bid at the auction. McCarter also engaged a well-known art dealer and friend, Richard Gray, to lend his experience to the bidding process.

McCarter and Field Museum curator of paleobotany and museum vice president Peter Crane went to New York in early October to attend the auction. McCarter, Crane, and Gray would be hidden in a private room above the auction floor with Gray making bids for the museum by telephone. The room was kept dark so no one could see who was in it. The Field would be competing with nineteen other bidders, including at least two other museums: the Smithsonian (also bidding in secret) and the North Carolina Museum of Natural Sciences (openly bidding from the main floor). The early estimates for SUE had predicted a price in excess of $1 million, but other estimates were as high as $12 million. The Field Museum and the Smithsonian both chose to bid in secret because they thought that if it were known in advance that they were bidders, it would drive the price up. It is unknown whether this strategy did much to stem the intense bidding once it began on the Sotheby's auction floor.

On the afternoon of October 4, 1997, Sotheby's auctioneer David Redden stepped up to the podium and started the bidding at $500,000. The bids progressed quickly, soaring in $100,000 increments and quickly surpassing the million-dollar mark. After $2.5 million, the Smithsonian dropped out, and after $7.2 million, the North Carolina Museum of Natural Sciences dropped out. Then Florida real estate baron Jay Kislak made a bid for $7.3 million. The Field countered with a bid of $7.4 million. Kislak bumped his bid to $7.5 million, and Crane told Gray that was the museum's prearranged limit. Gray urged Crane and McCarter to make one more bid. After a brief discussion, they agreed to let Gray make one last bid of $7.6 million. And that did it. Going once, going twice, and sold! We had it! With Sotheby's 10 percent commission added, the total price tag came to a whopping $8,362,500 for the yet mostly unprepared skeleton. The museum had been bidding against many others on the floor of Sotheby's including several secret bidders. One was a wealthy private collector who wanted SUE as an adornment for his spacious living room. Another was a major Las Vegas gambling casino. There were even rumors that the pop singer Michael Jackson had been one of the bidders. But in the end, the Field Museum won the day, and we were elated. The Field Museum had been successful in securing the world's most famous dinosaur for research and public education in perpetuity.

Once we had title to SUE at the Field Museum, the real work began. First, we had to move the bones from Sotheby's in New York to the Field Museum in Chicago. The Chicago company of Pickens-Kane Moving and Storage, specialists in fine arts moving, agreed to move the bones of SUE to Chicago pro bono. Bill Simpson and another Field Museum preparator, Steve McCarroll, went to New York to oversee the packing. The more than five tons of bones, plaster jackets, and rock were loaded into a single large semi with two levels (double floor space). The journey took the late afternoon and all of the night. A car with an armed guard followed behind the truck all the way to Chicago. Once the truck arrived in the parking lot of the museum the next morning, the enormity of what we were in for really began to sink in. It was a media circus. There were television reporters and camera crews from all over the area waiting to catch a glimpse of the world's most famous dinosaur. The field jackets and bones were unloaded, and Simpson took charge. Although the BHI had already put about 4,000 hours of preparation time into the specimen, that represented only a fraction of what was necessary to complete the job. Most of the bone was still encased in rock, and many bones were broken and in need of reassembly. We welcomed SUE to the museum with a short exhibit called *SUE Uncrated*, which drew tens of thousands of people. Before final preparation, we shipped the skull to Rocketdyne, a facility in Los Angeles that had a CT machine big enough to scan the entire SUE skull encased in a large protective crate. There it was scanned for details about the internal anatomy, which would give the preparation team an orientation to delicate structures inside. Then the specimen was shipped back to Chicago. Over the next two years, twelve preparators under the supervision of Simpson would put in another 30,000 hours of preparation to get the skeleton ready for study and mounting. The front half would be prepared at the Field Museum in Chicago while the back half would be prepared at a second Field Museum preparation lab set up at Disney World in Orlando, Florida. It was a monumental project. Simpson helped design the lab in Florida as well as modify the lab in Chicago to accommodate the project.

People came from all over the country to get an early glimpse of SUE's skeleton. They could follow the progress on the dinosaur's preparation and assembly because there were special preparation labs in both Chicago and Orlando with a glass wall for the public to view the progress. There were even cameras set up in the lab in Chicago with live feeds to the Internet so people could view the activity online. From school groups to local politicians, from scientists to media, the word of SUE's impending debut was spreading fast. After we had prepared and reassembled the foot of SUE in Chicago, movie director Steven Spielberg came to see it. He was two-thirds of the way through his blockbuster *Jurassic Park* movie trilogy, whose star was a *T. rex* realistically brought to life by his talented cinematic team. He had portrayed the tyrannosaur as a ferocious giant of slightly exaggerated proportions. Upon seeing in our preparation lab the articulated foot of the largest known *T. rex* ever found, his first words were something to the effect of "I thought it would be bigger." Some people are harder to please than others.

Just as things started to be going smoothly, a new point of tension surrounding the specimen surfaced. Peter Larson claimed to own the rights to the name SUE. (It seemed like nothing was easy with this fossil!) Negotiations on the name ownership broke down after a while, and the museum finally announced it would stop using the name SUE and launch a nationwide contest for elementary school children called "Name the *T. rex*." Negotiations between the museum and Larson had been taking time away from what should have been research, education, and promotional objectives for the museum and SUE. It seemed that the best way to move ahead on SUE would be to simply rename the fossil. To enter the contest, children would have to submit an essay on the fossil along with their suggested name. We were swamped with entries. Then, a few weeks after the contest was well under way, the dispute between Larson and the museum evaporated. Susan Hendrickson and Bob Bakker had reasoned with Larson on behalf of the Field Museum, and he subsequently gave all rights to the SUE name to the museum. By that time the museum couldn't simply declare the contest null and void because there had already been 6,000 entries! So we selected the most frequently submitted name "Dakota" as the best name, and we picked the three best essays among the pool of contestants that had suggested that name. Each of the three winners received a new desktop computer as a prize. After announcing the winning contestants, the museum then announced that it had decided to keep the name SUE because we now had the rights to the name and it was the name the public most readily associated with this specimen. And as it turned out, the rights to the name Dakota were already owned by at least one other company.

The next step in bringing SUE to the world was mounting the skeleton for public exhibition. The vertebrate paleontology curators (Flynn, Bolt, Rieppel, and I) and Simpson had all discussed various options for doing this after SUE had arrived in Chicago. There was a unanimous agreement that the real skeleton would be used for the exhibit mount. Some museums might have preferred to mount a cast and maintain the real bones behind the scenes in cabinets for easy research access, but the Field Museum has a practice of displaying the original specimens in its exhibits. So the question became who would mount the prepared bones in a way that would make an impressive exhibit while retaining sufficient access for scientific study? We sent the project description out for bids and received several responses. There was an extreme range of price quotes for the job, from $200,000 to $2 million. In reviewing all the submitted proposals, and in consideration of the uniqueness of the SUE skeleton, we ended up choosing the most costly of the proposals: that of Phil Fraley, formerly from the American Museum in New York. It was clear that his plan was the best and most fitting for the most famous dinosaur of all time. I had remembered Fraley from my days as a graduate student in the American Museum. He had been project manager of exhibitions there, responsible for mounting specimens in the exhibit halls. At the end of each day when I left the museum to go home, I had to pass through part of his mount shop on the way outside. Occasionally I had stopped to chat with him or his staff, and I knew he and his team were talented and reliable. Once we told him of our decision to hire him for SUE, he asked the American Museum for a leave of absence to do the project. They refused, so he resigned from the museum and started his own company, Phil Fraley Productions, launched with the mounting of SUE for the Field Museum.

 The Fraley mount was a monumental piece of engineering genius. He had a team of welders, metalworkers, engineers, and artists whose job was to design a mount that would hold the bones as an attractive assembled skeleton in a way that would require no drilling of any of the bones, no permanent glues, and no anchors to the bone. It had to allow for any single bone to be removed for study whenever necessary. It was a mounting job of unprecedented scale that was a challenge even for Fraley and his experienced team. Each bone was to be like a diamond in a setting, cradled in its own forged metal bracket. Most of the metalwork was done in sections at the Fraley shop in New Jersey. Each individual bone of SUE, once prepared, had to be carefully shipped to New Jersey for sizing and shaping of the black-colored metal structures. It was determined that the real skull would be too heavy for the mount, so it was to be displayed in an exhibit case on the balcony above the SUE mount, and the skull on the mounted skeleton would be a much lighter hollow cast of the original. Eventually the entire skeleton, together with the cast of the skull, was mounted in New Jersey in Fraley's shop. Once the complete mount passed Fraley's strict standards, it was disassembled and shipped back to Chicago six weeks before the scheduled unveiling. A large 20-foot-high temporary wall went up in the main hall around the place where SUE would be exhibited, and Fraley and his crew worked for long hours through the day and night putting everything together. They had asked for six weeks to do this, but once Fraley and his crew started the last phase in Chicago, they became caught up in the excitement of it all. They finished in just over two weeks. So for the next four weeks, the complete mounted skeleton of SUE remained hidden behind the temporary wall, waiting for the scheduled unveiling. Prior to that day, the museum did not want to reveal details of SUE's pose or the beauty of Fraley's work.

Part of the museum program surrounding SUE included continual research on the skeleton. We did not at the time have a curator of dinosaurs, so in January 1998 we hired a postdoctoral researcher associate to focus on SUE, Chris Brochu.[5] Ironically, Chris was born on August 12, the same date that SUE was discovered by Susan Hendrickson. He was a reptile specialist who worked mostly on fossil and living crocodiles and alligators. He also had a strong knowledge of dinosaurs and was expert at using CT scanning data of the sort we were getting from Rocketdyne for SUE. Chris studied SUE all through the preparation process, and the result was a 138-page monograph about SUE's anatomy that was later published by the Society of Vertebrate Paleontology.

The museum chose May 17, 2000, for the official unveiling of SUE in part because it was a free admission day and we wanted to share the excitement of the day with as many people as possible. The mayor of Chicago, Richard M. Daley, was there, as were hundreds of members of the media and thousands of other visitors. Peter Larson and Susan Hendrickson were also there to see SUE's grand opening in Chicago. Museum attendance on the day of the unveiling was about 10,000, but the TV coverage on more than fifteen stations went to hundreds of millions of people across the world. Dozens of newspaper reporters from all over the world were also there. It was a huge day for the Field Museum. SUE was an immediate success. Two days later President Bill Clinton and several U.S. senators visited the museum to see the famous fossil. Now that SUE was in a major natural history museum, it would be cared for and fully accessible to the scientific community in perpetuity.

The immediate success of SUE in Chicago attracted donors of varying capacity to the museum, including one who thought the Field Museum should have its own curator of dinosaurs. This anonymous donor contributed an amount that would not only endow a curatorship, but also a fossil preparator to go along with it. In 2001 Pete Makovicky was hired as curator of dinosaurs. Pete was a student at the American Museum, much like I had been before coming to Chicago.[6] He had grown up in Denmark, and as far back as he could remember he had always been interested in dinosaurs. He competed with scores of other PhD applicants for the job, but he won the competition hands down. His fieldwork has covered regions all over the world, from Mongolia to South America, from the western United States to Antarctica.

Pete wrote papers revising the weight estimates for SUE in life (from 7 tons to 10 tons) and documenting some of its ailments in life (arthritis and various infections). He calculated an age of twenty-eight for SUE at the time of death by counting annual growth lines in several of the bones. This indicated a phenomenal growth rate of over five pounds per day during its prime growth years. This animal had to eat a lot of other dinosaurs to sustain such an incredible rate of growth and must have been a terror in the Cretaceous community!

The rapid growth and gargantuan size came at the cost of agility. The muscles of the hips and legs were massive, but they functioned mainly to support and balance the animal's huge mass. Nevertheless, due to the animal's great size, it still could have achieved a peak running speed of between 10 and 25 miles per hour.

Pete Makovicky also oversaw a project to create the first digital *T. rex*. This was a lengthy process requiring CT scanning, laser scanning, and white light scanning of all of SUE's bones, both as individual bones and as an assembled mounted skeleton. All of the scanning data were combined to provide the first 3-D digital atlas of a *T. rex*. This digital file can now be used for 3-D printing of any of SUE's bones to exact scale at any size. Three-D "prints" are three-dimensional copies resembling casts that can be made of plastic, plaster, metal, or even food (chocolate SUE skull, anyone?) Researchers do not need to come to Chicago to take measurements on the SUE bones. The digital files can be sent anywhere around the world electronically, and institutions with their own 3-D digital printers can even create their own exact replicas of SUE bones for study. One-eighth size 3-D skulls of the specimen were produced in plaster for sale in the museum's store so anyone can have a miniature skull of SUE at exact scale (page 155).

The saga of SUE is a controversial story of discovery and passion, justice and injustice,[7] and finally groundbreaking research and international public exhibition. As a unique fossil now fully prepared and accessible, it can be examined in detail by the world's scientists. According to Pete Makovicky, SUE has produced more information about *T. rex* than any other specimen over the last century. More than fifty scientific technical publications have been published on this skeleton as of 2016, as well as hundreds of articles in the popular press. By late 2016, over 24 million visitors had seen SUE at the museum. In addition, we made casts of SUE for traveling exhibits that by 2016 had toured in more than ten countries around the world and were viewed by about 10 million more people. In coming to Chicago, SUE ended up in exactly the right place. SUE has been shared with the world in a way that few other institutions in the world could match. In the process, this *T. rex* has become the museum's most recognizable icon and one of Chicago's as well.

My experience with SUE gave me a new appreciation for the underlying politics, emotions, and ethical controversies that exist in paleontology, especially when it comes to dinosaurs. And Peter Larson of the BHI? The controversy still continues. In 2015 the South Dakota legislature overwhelmingly passed a resolution (103 to 2) to request that President Obama issue a formal pardon for Larson. At the time I finished this book, no decision had yet been made in Washington.

On August 12, 1990, Sue Hendrickson found SUE the *T. rex* (the horizontal brown layer to the right of her hand) weathering out of the side of the cliff on Maurice Williams's ranch. The skeleton was not much to look at yet, and at the time she had little idea how complete or how famous this fossil would become.

After Sue Hendrickson showed Pete Larson what she had discovered, Pete and his brother Neil began removing 30 feet of overlying rock from the bone bed containing SUE's skeleton. The cliff face where SUE's bones had partly weathered out was covered with a plaster jacket and a blue tarp for protection from falling rubble.

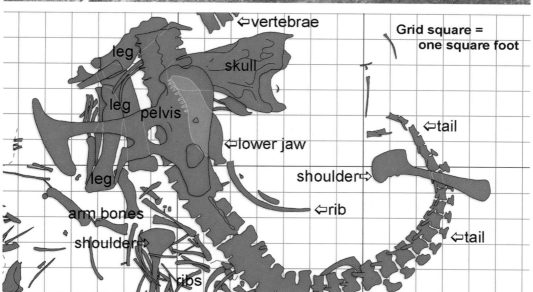

After weeks of digging in temperatures often exceeding 100°F, the Black Hills Institute of Geological Research (BHI) crew exposed the rest of SUE's skeleton. (*Top*) Showing the exposed bones still in the ground. Peter Larson is in the blue shirt behind Sue Hendrickson. (*Bottom*) Site map that the BHI drew to document positions of the bones in the ground before excavation.

One of the blocks of sandstone containing the bones of SUE after encasing it in aluminum foil, burlap, plaster, and two-by-fours. About 20,000 pounds of such blocks were trucked off of Maurice Williams's ranch to the BHI's facility in Hill City for the beginning of preparation.

Vince Santucci, "the pistol-packing paleontologist" of the National Park Service (so dubbed by members of the press and the superintendent of the Petrified Forest National Park). In the mid-1980s, he began a long investigation into the illegal collecting of fossils from public land, particularly from the Badlands National Park. To his dismay, the Department of Justice later rolled his case into the ill-fated prosecution involving SUE.

(*Top*) Hill City protesters outside of the Black Hills Institute in May 1992 protesting against the federal government's seizure of SUE. (*Bottom*) Army National Guard soldiers loading the confiscated boxes of SUE bones onto military transport trucks. Years later the government returned the bones to ranch owner Maurice Williams, who consigned them to Sotheby's auction house in New York.

The hammer came down at Sotheby's on October 4, 1997, making the Field Museum the new owner of SUE the *T. rex*. The winning bid was $7.6 million ($8,362,500 with Sotheby's commission). Standing behind the skull of SUE from left to right: John McCarter (then president of the Field Museum) Diana Brooks (then president of Sotheby's), Peter Crane (then director of the Field Museum), Richard Gray (Chicago art dealer who bid on behalf of Field Museum), and David Redden (the auctioneer).

Bob Masek, one of the twelve Field Museum fossil preparators working under the supervision of chief preparator Bill Simpson on SUE the *T. rex*. Here he is shown working on the underside of the skull in 1998.

Susan Hendrickson and I in February 1999, admiring the fully prepared right foot of SUE the *T. rex*. It had just been returned from the museum's fossil preparation team at Disney World, and Susan was visiting the Field Museum to help promote the upcoming SUE unveiling. A few days later, blockbuster movie direct Steven Spielberg of *Jurassic Park* fame also stopped by to see the foot.

The author (*right*) talking with Phil Fraley (*second from right*) and the rest of his SUE assembly team on April 19, 2000, only a few weeks before the public debut. All assembly work was done behind a temporary 20-foot wall built around the dinosaur in the main hall of the museum.

The official unveiling of SUE at the Field Museum on May 17, 2000. Nearly 10,000 people came to the museum that day, and in the sixteen years since then, over 25 million visitors have come to see the iconic skeleton.

Distinguished visitors in to see SUE the *T. rex* during opening week in May 2000. *From right to left:* President Bill Clinton, museum president John McCarter, U.S. Senator Dick Durbin (mostly hidden), and U.S. Senator Tom Harkin.

Field Museum's curator of dinosaur paleontology since 2001, Pete Makovicky, taking measurements on the back of SUE's skull.

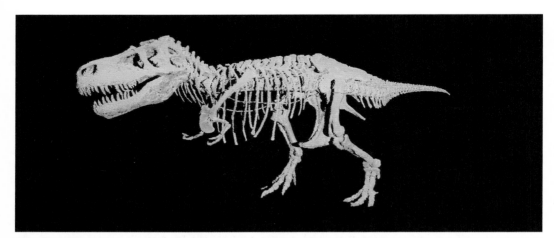
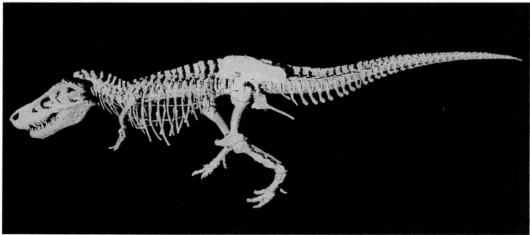

Surface scans of SUE's entire skeleton in 3-D made by the Forensic Services Division of the Chicago Police Department in 2009, under the guidance of Pete Makovicky. These data files, combined with CT scans of bone casts, have enabled the museum to reproduce any part of Sue's skeleton at any size via digital printing (see next figure).

Bill Simpson, the Field Museum's chief fossil preparator during the preparation of Sue, holding a 3-D digital print of SUE's skull at one-twelfth scale. "Prints" like this one (resembling precise 3-D casts) are made of polymerized gypsum with superglue and are produced from the scanning files generated from the SUE skeleton (see previous figure).

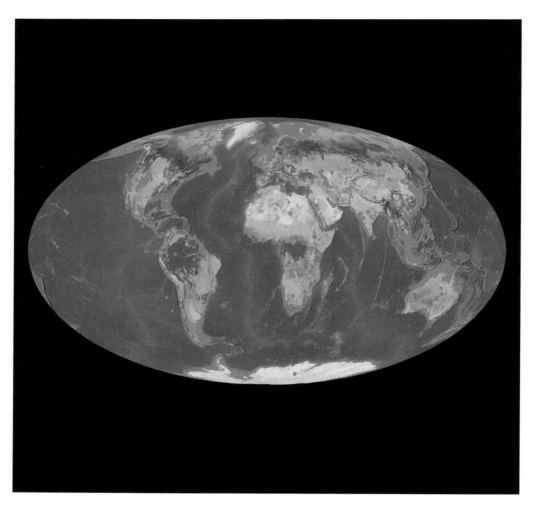

Field sites for the twenty-seven curators of the Field Museum in 2012 (*red dots*). I came to appreciate the diverse expertise and global reach of the collective curatorial staff. We all shared a formidable mission: to explore the diversity and history of the Earth and its inhabitants.

7

Adventures of My Curatorial Colleagues from the Field

Over time in Chicago, I got to know all of my fellow curators and something about their different research interests. I came to feel a sense of camaraderie with regard to our collective mission, curiosity, and experiences. The resulting synergy added depth to my own professional development. The Field Museum maintains a diverse range of research scientists. They explore Earth from the deep ocean trenches to the high mountain regions of the Himalayas, sometimes at risk to their own safety. They record the biodiversity of groups ranging from ants to dinosaurs and human culture ranging from Stone Age Europe to modern-day Polynesia. They decipher evolutionary history and develop strategies for environmental conservation. Maintaining such a diversity of expertise over the decades has given the museum the flexibility to respond to the ever-changing interests of society. You cannot predict which specializations or which regions of the world will lead to the greatest scientific successes of the future; but if you hire the most creative people who are passionate about their work, great and interesting discoveries will be made and novel applications will be found.

One inherent strength of museum research is in its broad interdisciplinary perspective. We combine fields such as geology and biology, or biology and anthropology, to look at larger questions. This is facilitated by the diversity of scientific expertise within the museum, and also by the comparatively small size of the organization. In 2014, for example, there were close to a thousand research professors on staff at the University of Chicago, but only twenty-one curators at the Field Museum.[1] Because the research faculty all operate within the same building, it is often easier to cross departmental lines both physically and intellectually. We attend each other's seminars, use each other's collections, and occasionally collaborate with one another on educational initiatives. We combine different types of data and methods for an integrative approach to some of the largest questions about nature and human culture. In spite of the many different areas of curatorial focus in the museum, we all share a passionate interest in better understanding natural history and cultural diversity. My Field Museum colleagues illustrate the breadth of research that a major natural history museum covers.

One of the world's leading experts on the ecology, evolution, and diversity of mushrooms, and longtime colleague of mine, is Greg Mueller.[2] He is also one of the few mushroom conservation specialists in the world. Mushrooms and other fungi are critical members of Earth's natural recycling system as the main decomposers of plant compounds such as cellulose. Without them we would be buried miles deep in dead leaves and logs. Mushrooms have an interesting evolutionary history and are more closely related to animals (including humans) than they are to plants. It has been estimated that there are over 100,000 different species of mushrooms on Earth, and fewer than 20,000 of these have been described and named so far by scientists. Greg is one of the scientists chipping away at the backlog of unidentified mushroom species and is a patron of the fungus among us.

About half of all mushroom species are edible, most of the rest can make you sick, and about 100 of them are deadly. In the United States, there are over 2,500 cases of mushroom poisoning reported each year, with 500 resulting in serious illness or death. Being the region's mushroom specialist keeps Greg on call with local hospitals. Whenever there is a case of mushroom poisoning in the Chicago region, Greg is the one who is asked to identify the ingested bits of mushroom and help physicians provide the most effective course of treatment. In the summer of 2007, I went on a field trip with him in Lake Forest, Illinois, along with a group of museum supporters and trustees anxious to learn something about their local wild mushrooms. In the scope of about an hour, Greg found over twenty species. Greg's wife, Betty Strack, is a technician who manages the scanning electron microscope laboratory at the Field Museum. When Greg collects edible mushrooms to bring home for dinner, Betty insists on re-identifying every mushroom herself to make sure it is safe to eat. You might think that this could be a blow to the ego of one of the world's preeminent mushroom specialists, but Greg says he appreciates the double-checking. In 2009 Greg left the Field Museum to become vice president of science at the Chicago Botanic Garden.

Field Museum botanist Rick Ree[3] studies flowering plants, of which there are estimated to be over 400,000 species. Some of his recent collaborative work includes the identification of plant species with possible anti-tuberculosis compounds, but his deepest interests are in evolution and biodiversity. He seeks to understand how geography, insect pollinators, birds, and plants are all linked through the process of coevolution. Nature is an ever-evolving network of interdependent organisms.

Rick does much of his fieldwork in the high mountain regions of China and India, which reach nearly 20,000 feet above sea level. It is surprisingly a hotspot of plant diversity rivaling the tropics. He has a plant survey program with colleagues from the Chinese Academy of Science in which he has collected thousands of flowering plants. Seeing some of his collections from 2006 is what first drew me into the museum's impressive herbarium sheet collection (e.g., page 181). Each herbarium sheet has an entire plant that has been dried, pressed, and mounted on thick archival paper. These beautiful sheets document the world's plant diversity, from the tropics to the polar regions. The Field Museum is currently working to digitize its collection of nearly 3 million herbarium sheets representing over 100,000 species as a source for online identification guides. Part of this effort, including high-resolution images of herbarium sheets, can already be accessed (search for "Neotropical Herbarium Specimens: The Field Museum").

Lichens are composite organisms consisting partly of algae or cyanobacteria and partly of fungus. The Field Museum's Thorsten Lumbsch[4] is one of the world's experts on this very strange group. The fungus component of lichens provides structure and shape for the organism, while the cells of algae or cyanobacteria allow it to produce its own energy (food) through photosynthesis. These peculiar organisms are poorly known yet are represented by over 20,000 living species, many of which resemble little more than dried paint crust. They have fascinating and unusual qualities. Scientists believe that some lichen species do not age and may be effectively immortal (although they can be killed by outside forces). They are also extremely adaptable, with species found from the Sahara desert to Antarctica, occupying every possible terrestrial environment on Earth. Some species are near indestructible. In 2008 a group of lichens was sent to the International Space Station and kept on the outside hull for a year and a half. After more than a year of exposure to the vacuum of space, cosmic rays, and extreme temperature variations, they were returned to Earth and 71 percent of them started growing again as though nothing had happened.

Thorsten's research on lichens focuses on their evolutionary history, origin, and geographic distribution. He is very productive, having published over four hundred scientific papers and described and named over a hundred new species of lichens before the age of fifty. He is also responsible for a car door being cataloged into the museum's botany collection. While he and adjunct curator Robert

Lücking were in Puerto Rico for a scientific meeting, they discovered and collected a rusty car door with an amazing diversity of lichen species growing on it (page 182). They estimate that it contained nearly one hundred different species! The door became a centerpiece of a Field Museum exhibit on lichens that opened in 2014.

One of the world's leading experts on South American plants and on the effects of El Niño is Michael Dillon, another longtime colleague of mine.[5] He was curator of botany at the Field Museum for thirty years. Dillon, as he prefers to be called, is a bit of a legend. He has described and named dozens of new plants from Peru and Chile, and there have been over two dozen plant species named in his honor by colleagues. He is the only scientist that I know personally who has a scientific journal series named after him. (*Dilloniana*, a Peruvian botanical journal published in Spanish.) In 2012 a Peruvian institute for environmental and cultural conservation was named in his honor, El Instituto Científico Michael Owen Dillon, or IMOD (http://www.imod.org.pe/). In Peru there is no botanist held in higher esteem than "Dillon of the Andes," as he is often called by his colleagues.

One of Dillon's more swashbuckling experiences in the field includes performing briefly in a Mexican rodeo in 1979 in order to gain access to a cooperative (*ejido*) in northern Coahuila. There were many interesting endemic species of plants lurking in this large community-run ranch area, and he wanted access to collect there. But the ranch hands were in no hurry to open the gate and would not respond to Dillon's pleas to let him in. One day they set up a local rodeo, and Dillon decided to come and watch. After being teased by the ranch hands for nearly two hours to get into the corral and ride one of the excited quarter-ton cows, he finally said, "The hell with it," and jumped into the ring. He says that it was one of the most terrifying experiences of his life; but there was no going back once he saw the gate open and heard someone slap the cow on the butt. He was not on the bucking bovine for very long before the animal threw him off and painfully trampled him, leaving him badly bruised. But the Mexican crowd loved it and roared with applause! And after this great show of bravado, the ranch hands shared a cigarette with Dillon as one of their new comrades and then let him on the property to collect all of the plants he wanted.

For the last twelve years, ornithologist John Bates[6] and his colleagues have been seeking to study and protect the diversity of birds in the Albertine Rift Mountains of central Africa. The region has been forgotten by most of the world as its civil society has been decimated by genocide in Rwanda and a Congolese civil war that has raged for years. Special measures to protect people's lives must be taken in order conduct research there. This means being accompanied with machine-gun-bearing guards while checking mist nets used to collect birds in Kahuzi-Biega National Park and hiring local militia groups for protection in other areas. Working in the region depends on building trust with and showing respect for his Congolese colleagues. John's life and the lives of his field crew depend on it. For them, biodiversity research and fieldwork have to happen even in areas of human tragedy, because humans are both the problem and the solution in these regions. He trains graduate students in the Congo as well as in other regions to create local legacies of expertise on biodiversity that will continue long after he stops doing fieldwork in those regions.

Another important way that the bird collection grows, as well as other zoological collections, is through salvage programs. Some two hundred species of birds regularly pass through Chicago during the spring and fall migration seasons. After a weary night of migrating over the expanse of Lake Michigan, they seek a place to rest and are attracted by the lights in the downtown buildings. They cannot see the clear glass windowpanes and fly into them, which is usually fatal. During the peak of spring and fall migration, a new dawn can mean that the city sidewalks below large buildings are littered with dead and badly injured birds. More than ten thousand birds are killed each year this way. Museum staff and volunteers, led primarily by John's collection manager Dave Willard, work with outside groups in a citizen-science approach to salvage the dead birds for science. Some 4,000 to 5,000 dead birds are added to the museum collection each year this way. Many of these specimens are skeletonized in the dermestid beetle room of the Field Museum, and this program has helped build one of the world's largest collections of bird skeletons.

Since John and his wife, Shannon Hackett (mentioned later in this chapter), both joined the Field Museum as curators of the Bird Division in 1995, the collection has grown by over 90,000 specimens.[7] It includes more than 90 percent of the world's 10,000 species of birds and is one of the four best bird collections in the world.

After being at the museum for many years, I was amazed to find out that we have pieces of the planet Mars in our collection, in the form of Martian meteorites. I became familiar with this when Meenakshi Wadhwa,[8] or Mini as she is called by her friends, was the Field Museum's curator of meteoritics from 1995 to 2006. She is one of the world's leading meteorite scientists, and these fragments from the red planet were a special expertise of hers. Martian meteorites are extremely rare and identified by the gases trapped inside them, which match the atmosphere of Mars. We know the exact composition of the atmosphere on Mars thanks to analysis by the Viking spacecraft that landed there in 1976. Of the more than 61,000 meteorites that have been found so far on Earth, only 132 have been identified as Martian in origin as of March 2014. How did pieces of Mars find their way to Earth? Millions of years ago, a large asteroid struck the red planet with enough force and mass to blow large pieces of it into space as meteoroids. After traveling tens of millions to hundreds of millions of miles along the right trajectory, the Martian meteoroids entered Earth's atmosphere to become meteors. Those Martian meteors that didn't burn up on their descent hit Earth as meteorites (a *meteoroid* does not become a *meteor* until it enters Earth's atmosphere, and a *meteor* does not become a *meteorite* until it lands on Earth).

One of Meenakshi's favorite field sites is Antarctica, where she has been on two major expeditions. There you can even look for meteorites by helicopter, on a snowmobile, and on foot. More than 95 percent of the continent's surface is covered with snow and ice averaging 7,000 feet in thickness. These thick sheets move very slowly from the interior of the continent out toward the coastline. They have been accumulating ice for hundreds of thousands of years, and during that time meteorites have become embedded and been carried along. In some areas the ice gets blocked by mountain ranges, and the wind can ablate the ice, exposing thousands of years' worth of accumulated meteorites. You can sometimes find hundreds of meteorites in an area the size of a football field. The dark rocks you spot on the surface of the bluish-white ice fields are not obscured by vegetation and thus are easy to spot. In most areas of Antarctica, any rock you see on the surface of the ice is likely to be from space because there is no other source of rock. That is not to say collecting in the Antarctic is like a walk in the park. Sleeping in tents for a month or more when the windchill outside is −70°F can be a character-building experience. Meenakshi has made two trips there, each time spending six to seven weeks sleeping in a tent. Such work is not for the faint of heart.

In Chicago on March 26, 2003, something happened that most meteoriticists only dream about. A meteorite fall including hundreds of meteors was observed in the southern Chicago suburb of Park Forest, Illinois. This was the most densely populated region to be hit by a meteorite shower in modern times, only thirty-five miles from the city of Chicago. The fireball from the meteoroid hitting Earth's atmosphere was visible from Illinois, Indiana, Michigan, Wisconsin, Ohio, and Missouri. It is estimated that when the meteoroid hit Earth's atmosphere and exploded, it was traveling 44,000 miles per hour. The next morning, Mini packed up the museum truck and drove to Park Forest to see the fall site and to look for meteorites. By the time she got there, meteorite collectors and meteorite dealers had descended on the area in droves, creating a circus-like atmosphere. Some were there to look for meteorites themselves, and others were there to buy meteorites from the fortunate ones who found them. Some had fallen in the woods; others had fallen in parking lots or roadways. The dealers would actually pay extra for meteorites that had paint on them from hitting the yellow center line on an asphalt road. One meteorite crashed through the roof of a Park Forest resident's house, leaving a four-inch hole. At first this might seem unfortunate, but not when it comes to meteorites. The home owner was able to sell the section of roof with the hole in it along with the meteorite for enough money to pay for an entire new roof and then some. The meteorite damage was like winning the cosmic lottery for the home owner. A few weeks later, Meenakshi was able to obtain some of the Park Forest meteorite for the Field Museum's collection. Most meteorites have deteriorated by the time they are discovered because they begin to weather and alter as soon as they land on Earth. Mini was able to preserve this one in a pristine state.

In 2006 Mini left the Field Museum to become the director of the Center for Meteoritic Studies at Arizona State University. After she left, it wasn't long before we realized that with one of the world's largest, most important collections of meteorites, we were badly in need of a meteoritics curator. Unfortunately, there was a hiring freeze on curatorial positions at the time. As senior vice president and head of Research and Collections at the time, I began looking into ways to address this problem. After receiving a good lead from a research associate in the Geology Department, I began discussions with a major donor and his foundation about the possibility of endowing a meteoritics program at the museum (discussed further in chapter 9). In 2009, after several months of proposals and negotiations, he and the foundation agreed to endow a curator of meteoritics, a collection manager of meteoritics, and annual operating expenses for a meteoritics institute at the museum. It couldn't have come at a better time. Now we could hire a new curator without impacting the annual budget of the museum, which was at the time was in a very challenging phase.

In 2010, after a several-months-long search, we hired Philipp Heck[9] as the new curator of meteoritics. Phil has several different research specialties, including the study of fossilized meteorites found in sedimentary rocks more than 465 million years old. As you might imagine, this specialty went over well in a Geology Department that now consisted solely of paleontologists. Fossilized meteorites are extremely rare. Of the more than 50,000 meteorites that have been found on Earth so far, only about 100 of them are fossils. When you think about it, it is quite remarkable that we find any fossil meteorites at all given the relative scarcity of modern meteorites and the fact that most meteorites would not have been preserved as fossils. But there were times in Earth's past when it was bombarded with meteorites at a much greater rate than today. Around 470 million years ago, a large asteroid catastrophically collided with another body in the asteroid belt situated between the planets of Mars and Jupiter. This event put tens of millions of meteoroids into orbit around the sun. Many of these fragments ended up on a collision course with Earth. For several million years, there was a large surge in meteorite falls all over the planet that left a fossil record in rocks between 470 million and 465 million years old.

Phil's main specialty is research on material from space that predates the birth of our sun. He is searching for the parent stars of our solar system and piecing together the history of our galaxy. He is looking for the primordial substances from which all things were made. He and his colleagues have found nanodiamonds within meteorites from exploding stars dated to be 5.5 billion years old, which is a billion years older than Earth itself. He was part of the team that analyzed the first comet and interstellar dust brought back to Earth by NASA's Stardust mission. Phil has a great line that he often uses in his talks: "We are all stardust."

Before the age of dinosaurs, there was the age of an ancient reptilelike mammal relative called the dicynodonts. Ken Angielczyk[10] is one of the world's leading experts on this group and on the mass extinction event that decimated them, opening the door for the age of dinosaurs. Dicynodonts first appeared about 270 million years ago and rapidly became the most successful group of plant-eating animals of the time. There were hundreds of species of dicynodonts, ranging from some the size of a hamster weighing a few ounces to others that were the size of a hippo weighing over a ton. Some were burrowers; some were grazers. Based on life reconstructions hypothesized by paleontologists, most of them resembled something between an overweight American football lineman and a saber-toothed frog. During Permian times (270 to 252 million years before present), they were among the most common and geographically widespread of all land vertebrates. Their fossil remains are today present on every continent, although this is not so surprising given that during their existence as living animals, all of today's continents were connected together as the supercontinent Pangaea. By the Early Cretaceous (about 100 million years ago), dicynodonts appear to have become totally extinct.

Ken also studies the causes of and the recovery from the mass extinction event of 252 million years ago at the end of the Permian. This event, sometimes referred to as "the Great Dying," was the largest of all known mass extinctions in the geologic record, and it is estimated that it resulted in the extinction of 90 percent of all plant and animal species on the planet (including most of the dicynodonts). Rapid global warming is thought to be one of the factors that caused this extinction event. After the Great Dying, when Earth's biosphere started to recover, one of the major groups that came to prominence, filling the open niches left by dicynodonts, were the dinosaurs. They became the dominant terrestrial group of vertebrates for more than 135 million years until the end of the Cretaceous 66 million years ago, when dinosaurs succumbed to a large extinction event that wiped out around 75 percent of all plant and animal species living at the time. We can extrapolate lessons from the fossil record about the extreme effects of climate change, mass extinctions of life on this planet, and how the survival of humans as a species should not be taken for granted. We are but one of millions of species that have existed on this planet over the last several billion years. In the words of Stephen Jay Gould, from his book *Ever Since Darwin*, paleontology and geology have given us "the immensity of time and taught us how little of it our own species has occupied."

Not one to be claustrophobic, Janet Voight[11] is a marine biologist and curator of invertebrates who specializes in octopus biology and exploring the ocean abyss. She takes a deep-sea submersible called the

Alvin to the bottom of the Pacific at depths of more than a mile and a half deep. It can take hours to sink that far down in this vessel, the inside of which is not much bigger than the that of a compact car. So far Janet has made eight dives in the Alvin, which belongs to the U.S. Navy and is operated by the Woods Hole Oceanographic Institution (WHOI) for the national oceanographic community. The outside pressure at those depths is more than 250 times that of Earth's surface and is so great that it compresses an eight-ounce Styrofoam coffee cup down to something the size of a large thimble. (Janet gave me one of these from one of her dives as a souvenir that I keep on my bookcase at the museum to this day.) The spherical-shaped titanium chamber at the heart of the Alvin where Janet sits is kept at surface pressure and is the only thing keeping Janet from being crushed in an instant.

Janet is an explorer in every sense of the word. She studies the strange communities of organisms that have sprung up around deep-sea vents and have literally never seen the light of day. Completely devoid of sunlight and photosynthesis from plant life, these deep-water ecosystems are built completely around the heat and chemicals discharged from the undersea vents that look like small water volcanos. We have detailed knowledge of only a small sample of the biodiversity there (i.e., mostly the organisms that are not frightened off by the lights of the Alvin before coming into camera range). Some of the new species recently discovered from this environment are so unlike anything ever seen before on this planet that they have been classified in completely new orders of life. It has even been proposed by some scientists that life itself first originated on Earth around hydrothermal vents. The area of the deep ocean's surface is twice the area of the entire continental surface of the planet, yet we still know very little about it. In fact, as Janet often reminds me, we know more about the surface of the moon than we do about the bottom of the deep oceans. It is one of Earth's last great frontiers, and scientists like Janet risk their lives to help us better understand it.

Spider-Woman works at the Field Museum. At least that is what some call her. Curator of entomology Petra Sierwald[12] is one of the world's leading experts on spiders and millipedes. She handles tarantulas and giant millipedes like some people handle hamsters, and she is helping to decipher the evolutionary tree of life for these hard-to-love creatures. According to Petra, there are 45,000 described (named) species of spiders and another 13,000 described species of millipedes. God knows how many undescribed species of these animals still await discovery by scientists. Millipedes were among the first animals to have conquered land on this planet, with a fossil record going back at least 428 million years. In addition to her research, Petra has written some of the Field Museum's most successful federal grants for student training. These grants, from the National Science Foundation, have supported over forty-eight student internships at the museum during the last six years. These internships, involving hands-on experience with museum scientists, have encouraged many young students to aim for scientific careers themselves. There is a growing critical need for such educational efforts in the United States, and museum curators are ideally situated to provide them (discussed more in the last chapter).

Sometimes a better understanding of evolutionary relationships leads to seemingly unrelated applications and opportunities. Curator of birds Shannon Hackett[13] works on reconstructing the evolutionary tree of life for the 10,000 species of birds living today, using molecular data from DNA analysis. Her 2008 paper published in *Science* magazine with eighteen collaborators was a landmark study providing a tree of evolutionary relationships for all major groups of living birds. It is a standard reference used by all of the world's top bird specialists. As a result of that work, she and colleagues became involved in the Emerging Pathogens Project. This project studies the DNA of birds, small mammals, their parasites, and their pathogens (organisms that cause disease). The DNA she and her colleagues extract and analyze in Field Museum's Pritzker Laboratory for Molecular Systematics and Evolution helps scientists understand how diseases have evolved and what might happen as infectious organisms jump between animal species. By understanding the evolutionary relationships of the bird groups carrying the pathogens, we can trace the disease back to its original strain, to help fight it at its source.

The youngest curator at the museum at the time I wrote this chapter was Corrie Saux Moreau.[14] The compelling story of how she became a professional scientist was featured in a temporary Field Museum exhibition called *The Romance of Ants*, open from 2010 to 2012. It was an inspirational exhibit for young people aspiring to be scientists, and it even came with its own online comic book (http://www.fieldmuseum.org/sites/default/files/AntsComicbook.pdf). As the comic book relates, Corrie was born and raised in New Orleans, Louisiana. She was interested in nature and animals at a very early age, although there wasn't much in the way of natural wildlife in the city of New Orleans and the building where she lived did not allow pets. But in the cracks of the sidewalk around her home, she and her younger brother could always find ants. She began to identify the different kinds of ants, and she even did little experiments with ants like dropping crumbs to see how they would react. It's a story that reminded me of how my own childhood interest in natural history started while looking for fossils in my parent's gravel driveway.

Ants are as old as the dinosaurs, and they evolved from wasp-like ancestors more than 100 million years ago. They survived the mass extinction event at the end of the Cretaceous that wiped out as much as 75 percent of all living species on Earth, and they appear to have undergone an almost explosive speciation shortly after that. Today there are nearly 50,000 living ant species (of which only about 15,000 have been described). The total biomass of ants on the planet is enormous. According to Corrie, if you weighed all the ants on the planet, the mass would be equal to the biomass of all humans (7.4 billion people!). According to her blog site, there are approximately 321 pentillion ants alive on the planet today (321,000,000,000,000,000 ants).

Corrie maintained an interest in nature through high school and college, and eventually she began looking around for a university to pursue a PhD to study her favorite group of animals: ants. She applied to a number of universities, including Harvard, where the world-famous naturalist and ant specialist Edward O. Wilson was a professor. He was impressed enough with Corrie's application materials that he agreed to take her on as a student. Later Wilson, a two-time Pulitzer Prize winner, would include a story of Corrie in his book entitled *Letters to a Young Scientist* (2013).

By the end of Corrie's first year at Harvard, she already knew what she wanted to do for her thesis (as described in Wilson's book). Three leading ant specialists around the country had received a multimillion-dollar federal grant from the National Science Foundation to investigate the evolutionary relationships of all the subfamilies of ants using DNA sequencing. It was a huge undertaking, but if successful the project could set the stage for hundreds of studies by dozens of future scientists on the relationships, evolution, and ecology of one of Earth's most abundant animals. Corrie told Wilson of her plan to write to the three principal researchers on the project and ask for permission to work on just one of the twenty-one subfamilies of ants in the project. He agreed that it would be an idea worthy of a PhD and a good way for Corrie to establish connections to other experts in the field. A few weeks later, a disappointed Corrie came back into Wilson's office to tell him that the project leaders had turned her down. They were unwilling to include an untested graduate student on their team. Wilson was sympathetic but suggested she try something else.

A few days later, Corrie came back to Wilson's office, reinvigorated with a new thesis proposal. She had decided to take on the entire twenty-one subfamilies of ants herself (representing nearly 50,000 species). It was an amazingly ambitious goal that would have been too much for many students, but something about Corrie made Wilson instinctively agree to her proposal. And after a little over two years of long hours and focused research, she produced a cladistic pattern of evolutionary relationships for all twenty-one subfamilies. On April 7, 2006, her results were published as the cover article in *Science* magazine, and in 2007 she successfully defended her thesis on the phylogeny of ants before her PhD committee. The prestigious team with the multimillion-dollar federal grant published their results later that same year. Their phylogeny matched Corrie's almost identically. Since then, later studies have supported Corrie's version of the evolutionary tree of ants. Her story is a prime example of what courageous self-confidence and focused determination can do in science, and the type of passion and competitive drive curators often have that makes them successful.

Today Corrie's research also focuses on the role of ants in the evolution of flowering plants, and on their value as indicators of climate change. In addition to being a successful young curator herself, she founded the Women in Science group at the museum to encourage other women to pursue careers in the natural sciences.[15]

Many of my curatorial colleagues at the museum have also been anthropologists. Gary Feinman[16] is an archaeologist whose focus is on ancient civilizations and the evolution of early economic systems. His fieldwork is based primarily in Oaxaca, Mexico, and Shandong, China, which he conducts with his wife, Linda Nicholas, who is an adjunct curator of archaeology and ethnology at the museum. Gary is interested in why and how powerful states and empires rise and inevitably fall, such as the pre-Columbian city of Monte Albán in the Valley of Oaxaca, in southern Mexico, where he has spent many years doing fieldwork. There is much we could potentially learn from his research as Americans who today live in a modern-day empire of sorts. Perhaps by studying the downfall of past societies, we can prolong the success of our own.

Gary was one of the curators for the Field Museum's highly successful traveling exhibit *Chocolate*. This exhibit explores this substance from its origin in the cacao tree of the Central and South American tropical rain forests through time. Based on ancient residue in pottery dated at about 1200 BC, chocolate was first discovered and developed by the ancient Maya. To the later Aztec Empire, it was a luxury drink mainly for the elite. The cacao seeds were used for currency and for offerings to the gods. Chocolate eventually spread to Europe, and then to the rest of the world, where it has remained a cherished treat ever since. The exhibit premiered in 2002 at the Field Museum for eleven months before going on tour. Since then, it has been to twenty-four different states and Canada and has been licensed to Japan and Korea. It continues to generate revenue for the Field Museum, and according to the museum's head of Exhibitions, Jaap Hoogstraten, "This exhibit will travel forever. Who would want to live in a world without chocolate?"

The ancient Wari people formed a civilization in Peru that flourished from 500 to 1000 AD. They built Peru's first empire, which preceded the rise of the larger Inca Empire by several hundred years. Curator of archaeology Ryan Williams[17] and colleagues have been excavating a Wari site called Cerro Baúl for more than twenty years. This ancient city is more than 1,000 years old and lies about 8,000 feet above sea level. It was home to the elite members of the Wari Empire. In 2005 Ryan and his colleagues reported the discovery of the oldest large-scale brewery in the Andes at the site. The ancient brewery made a spicy beer-like drink called *chicha de molle* on a massive scale, and the facility could deliver the goods when hundreds of Wari thought it was *chicha* time. It could pump out the brew in batches of 475 gallons at a time. Ryan has been working in the Cerro Baúl site for more than twenty years, and in the process he has trained dozens of archaeology students there. He uses innovative new technology to explore what lies underground, such as a ground-penetrating radar machine on wheels. With this device, he has discovered underground tombs, caverns, and even buried ancient buildings.

The area of Cerro Baúl is located about 12 miles north of the city of Moquegua, a politically volatile place. In 2008 political troubles broke out between the Peruvian government and striking townsfolk of the region. There was a rising insurrection of workers and peasants there over redistribution of wealth, and there was even talk of revolution. The villagers had closed off the town of Moquegua near Ryan's field site, and violence had broken out. Many protesters and police were injured. Remarkably, one of the protesters brought down one of the Peruvian army's helicopters with a stone thrown by a sling. Ryan and the eleven members of his field party were trapped and cut off from supplies for several days. The Peruvian army was threatening to storm the region, which would have put Ryan and his team at even greater risk. As head of Collections and Research at the time, I prepared to implement the museum's emergency rescue plan for scientists threatened by such outbreaks. There are some situations in which even the U.S. State Department is not enough. The museum's general counsel, the CFO, and I began discussions with a professional "extraction team," which we have on retainer and whose job is to go in and rescue our people if the need should arise. Such an option has risks of its own, of course, and is only used as a last resort. Fortunately, we did not have to deploy the extraction team. At the last minute, locals secretly smuggled Ryan and his group out of the troubled area and to where Peruvian helicopters could fly them safely out of the region. Ryan's many years of careful networking with local people provided him and his field crew with a much-needed safety net.

From walking through the gates of Hades, to uncovering evidence of Stone Age love and piecing together the origin of social hierarchy in human society, archaeologist Bill Parkinson[18] has been part of some amazing discoveries. Bill is one of the lead archaeologists for the Diros Project. This project, organized by Anastasia Papathanasiou (Ephorate of Paleoanthropology and Speleology) under the direction of Giorgos Papathanassopoulos (Honorary Ephor of Antiquities), is excavating the awesome Alepotrypa Cave in southern Greece. This enormous chamber is the length of almost three football fields and filled with primitive stone jewelry, tools, and weapons. It also contains at least 170 human skeletons buried between 6000 BC and 3200 BC, and is one of the largest Stone Age burial sites in all of Europe. In 3200 BC an earthquake caused the entrance to the cave to collapse, sealing its contents for millennia. A number of Stone Age people were buried alive. The site was rediscovered in 1958 by two cave explorers, but it was not until the mid-1970s that archaeologists began to excavate the cave, which soon became known as the "Stone Age Pompeii." The site preserves one of the earliest records of large-scale human conflict, with 31 percent of the skeletons showing healed injuries of blunt trauma to the head, most likely inflicted by rocks or clubs. The cave houses a small lake, and its walls reveal ancient rituals involving fire. Excavation Director Giorgos Papathanassopoulos proposed that Stone Age people believed the Alepotrypa Cave to be an entrance to the kingdom of the dead (Hades' underworld in classical Greek mythology).

Although the Alepotrypa Cave site preserves one of the earliest records of large-scale human violence, the Diros Project also discovered a gentler side to Stone Age life. Just outside the cave entrance, the team found a double burial dating back to 3800 BC. It revealed two well-preserved adult skeletons, one male and one female, with arms and legs locked in embrace. They were both in their twenties at the time of death. As quoted by Bill Parkinson in a recent *National Geographic* article, "They're totally spooning . . . their arms are draped over each other, their legs are entwined. It's unmistakable." Love in the Stone Age? It is a powerful image that leads one to envision their last hours together and wonder what led to their deaths (page 192). The Alepotrypa Cave site documents an amazing window into prehistoric human culture and will continue to be explored by Bill and his colleagues for decades to come.

The collapse of the Soviet Union in 1989 opened up many new opportunities for Western archaeologists in Eastern Europe, a region that records humanity's emergence from the Stone Age. Bill and another team of students and colleagues have been excavating ancient sites in southeastern Hungary dating from 5500 BC to 4000 BC to follow this critical stage of humanity's social evolution. A momentous transition of human society called the Copper Age began around 5000 BC. Copper was the first metal used by humans in any quantity because of its widespread occurrence in nature as pure ingots that do not require smelting. It could easily be hammered into useful shapes. Learning how to use the metal was a major technological breakthrough resulting in the production of jewelry, tools, and weaponry. Even more important than the production of goods, the mining and processing of copper gave rise to new occupations, new social groups, and eventually to the first complex social hierarchies. It brought the earliest evidence of distinctions between rich and poor and between rulers and ruled, as well as recording the earliest beginnings of human urbanization. Once these kinds of societies developed, they spread like wildfire. Human culture went through a fundamental change, and today the majority of people on Earth live in urban settings. Bill asks the question of why humans began living together in dense urban settings, given that for many reasons they are not designed to do so. There are inherent problems that develop in dense populations with regard to sanitation, communicable diseases, and other factors. He also studies what has led to the failure and disappearance of many urban centers in the past. As part of his fieldwork over the last twenty years, Bill and his Hungarian colleague Attila Gyucha have trained over a hundred students in the archaeology of Hungary.

The Field Museum's longest serving curator, John Terrell,[19] has been curator of archaeology and anthropology for over forty-five years. His research explores human culture in the tropical regions of the Pacific, focusing on the diversity of human cultures and how they interact and evolve over time. One aspect of his work boils friendship down to a science and was presented in his recent book *A Talent for Friendship: Rediscovery of a Remarkable Trait*. In it he proposes that our capacity for friendship has evolved by way of Darwinian selection, just as arms, legs, and jaws have. He contends that the ability to make friends even with strangers is a characteristic that is uniquely derived for humans in nature (although through selective breeding, humans have also developed this trait in dogs and perhaps other domestic animals). He further suggests that developing a better understanding of the social evolution of friendship within *Homo sapiens* may enable us to "tinker with our evolutionary heritage in social . . . ways to help us achieve greater control over the future of our species."

Most of my anthropology colleagues have been cultural anthropologists focused on the study of past and present human societies. In contrast, Robert Martin[20] has been the museum's only physical anthropologist (a field that focuses on the evolution of primate biology and behavior). His research includes the controversial hominid species *Homo floresiensis*, nicknamed "the hobbit." Adults of this species were described as only three to three and a half feet tall, with a proportionately smaller brain size than *Homo sapiens*. Its fossils are from Indonesia in deposits as recent as 12,000 years old, making it the youngest known species of extinct hominid. It has been proposed that *Homo floresiensis* lived contemporaneously and interacted with *Homo sapiens* (modern humans). Bob's research suggests that this so-called "species," known by only a few assorted bones and a single well-preserved skull, is actually an adult *Homo sapiens* that was born with a birth defect called microcephaly. The somewhat heated controversy confirms why you often need multiple specimens to establish the validity of new species, especially with fossils.

Bob is also an authority on the evolution of reproductive and child-rearing practices of primates (including humans), a subject he covers in his recent book *How We Do It: The Evolution and Future of Human Reproduction*. Bob has a monthly blog site with *Psychology Today* in conjunction with this book.

Chapurukha Makokha Kusimba,[21] who likes to be called Chap, was the curator of African archaeology and ethnology at the Field Museum from 1994 to 2013. I visited Kenya with him in 2007 to explore collaborative opportunities between the National Museums of Kenya and the Field Museum. We were hoping to bring some of the Kenyan Museum's hominid fossils to Chicago for a temporary exhibit on human evolution. The trip was a transformative experience for me, giving me a deeper appreciation for the museum's diverse interests in Africa.

When he was five years old, Chap's family moved to Kenya to escape the political unrest and violence that began overtaking Uganda, starting with the worsening iron-handed rule of Milton Obote, which would later be followed by the even worse atrocities of Idi Amin, who took control in 1971 after a military coup. Uganda had been dubbed the "Pearl of Africa" by Winston Churchill in 1909, but by the time of the reign of Idi Amin (self-proclaimed "His Excellency President for Life"), it had become known as the "Killing Fields." Amin's brutal rule has been described by human rights groups as being responsible for killing a half million Ugandan people. Among the targeted individuals were students and intellectuals. Chap got out of Uganda none too soon.

As a transplanted youngster growing up on the slopes of Mount Elgon, Kenya, life for Chap was not only safer—it offered new interests. He explored caves in the hills that contained the remnants of past civilizations now unknown to the world. Some of these caves had ancient wall paintings. Occasionally Chap would collect an arrowhead or a stone tool and ask local elders about it. They would explain what they knew about the people who lived there in former times, and this made Chap want to learn more. He studied and did well in high school, and eventually became the first member of his family to go to college. At that time there were not many career opportunities for anthropologists in Africa. Then one day the opportunity of a lifetime materialized for Chap: an internship at the National Museums of Kenya. The invitation came from Richard Leakey, who was director of the institution. Leakey had been instrumental in building international collaborations between the National Museums of Kenya and the rest of the world. These collaborations effectively nationalized research in Kenya and made possible the training of several dozen Kenyan students in the field of anthropology. Prior to this, the field of Kenyan anthropology had been dominated by British scientists who had shown little inclination to train local scholars.

Chap developed a research program around the caves and artifacts that he had discovered as a child growing up in Kenya. He learned that the caves were where people hid from slavers in the nineteenth century. During that time, Swahili and Arab people from the coast would attack entire villages in the African interior and take people as prisoners to sell into the slave trade. The slave trade was to blame for the destruction of many East African societies, and some of Chap's own ancestors were taken long ago. Chap has spent decades studying the cultural evolution of the East African coast (sometimes referred to as the Swahili Coast). For an archaeologist, the prospect of trying to decipher the early history of civilization in Kenya is daunting. For at least two thousand years, the area appears to have been a mosaic of different cultures, religions, and economies. Civilizations grew and flourished for centuries as a result of trade along the coast, but then they were destroyed in the early sixteenth century by the slave trade. As a result, much of the cultural history of the region was lost. Chap's connections and family within Kenyan society give him a special advantage in reconstructing the history of the region through stories passed down from generation to generation. He has accumulated more than 800 hours of videotaped conversations with elders and other informants on the traditions, migratory histories, and life histories of people in East Africa. He is only beginning to understand the complex history of the region as someone who has made it his life's work to do so.

Over my years at the museum, I came to appreciate the collective mission of the curatorial staff. Although the research programs are diverse, covering many disciplines and geographic areas, the curators share a common sense of purpose. They all have a determination to turn their scientific curiosity into active, dynamic research programs and to disseminate knowledge of their discoveries through publishing, exhibition, and educational training. They are not afraid to take risks. They independently create their own social and professional networks to get things done, whether that means making personal connections to indigenous elders in remote villages, riding an angry cow in a rodeo, interacting with donors and volunteer organizations, or advising graduate students who will become the next generation of scientists. Each curator makes his or her own intellectual mark, and their cumulative impact is what makes a great scientific institution.

Field Museum curators of botany. (*Top*) Greg Mueller discovering a patch of the shaggy mane mushroom in the Caucus Mountains of Russia. (*Bottom left*) Rick Ree conducting a biodiversity inventory in the Hengduan Mountain region of China. He is holding a GPS device to record exact collecting site coordinates. (*Bottom right*) Thorsten Lumbsch collecting lichens just outside of Melbourne, Australia.

Two specimens of the flowering plant *Delphinium kamaonense* collected from the Himalayan region of China by curator of botany Rick Ree in 2006. They have been pressed, dried, and mounted on a herbarium sheet. The Field Museum's botany collection contains about 3 million herbarium sheets representing over 100,000 species of plants. They are all stored in sealed, environmentally controlled cabinets to last in perpetuity for future scientific research.

A rusty door from a 1984 Ford Bronco collected in Puerto Rico in 2014 by curator of botany Thorsten Lumbsch and adjunct curator of botany Robert Lücking. It was found with nearly 100 different species of lichens growing on its surface and is now part of the Field Museum's botany collection. Although many lichen species resemble nothing more than peeling paint from a distance, they are an important part of our biosphere and, no doubt, things of beauty in the eyes of the lichenologist.

(*Top left*) Michael O. Dillon, curator of botany at the Field Museum from 1979-2008, shown here in San Martín, Peru, examining a specimen of an umbrella tree (*Schefflera* sp.) in 1995. (*Top right*) An issue of the Peruvian botanical journal series named in his honor, *Dilloniana*. (*Bottom*) Website brochure from the Peruvian botanical institute named in his honor, the Instituto Científico Michael Owen Dillon (IMOD). Dillon is one of the most celebrated scientists in Peru.

Michael Dillon in 1979 photographed just after being thrown by an excited quarter-ton cow, during a rodeo in Coahuila, Mexico. He risked life and limb to bond with local ranch hands, who finally then allowed him access to collect plants on their ranch. Curators often go to great lengths to make connections with local people in order to establish productive field sites and effective safety networks.

(*Top left*) John Bates, curator of ornithology, helping a park guard learn how to prepare bird specimens in Malawi, Africa, in 2009. (*Top right*) Birds collected from central Africa by John and his colleagues that have been made into skins for the collection. (*Bottom*) Dozens of birds that were killed flying into buildings in Chicago and salvaged for the museum. Once in the museum, they were skeletonized in the dermestid beetle room and cataloged for the collection with the help of volunteer Kayleigh Kueffner shown here.

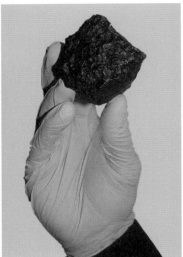

(*Top left*) Meenakshi Wadhwa, curator of meteoritics at the Field Museum from 1995 to 2006. She is holding a meteorite that fell in Park Forest, Illinois, on March 26, 2003. (*Top right*) A fragment of Mars in the Field Museum collection. (*Bottom*) Meenakshi standing outside of her tent in the Transantarctic Mountain region of central Antarctica on a trip to collect meteorites.

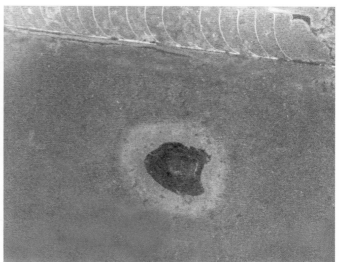

(*Top left*) Phil Heck, curator of meteoritics, holding a meteorite from the museum's collection. (*Top right*) Meteorite (middle of slab) and chambered squid-like cephalopod (top of slab) fossilized together in a 467-million-year-old limestone from Sweden. Slab is 8 by 10 inches in size. (*Bottom left*) Ken Angielczyk, curator of paleontology, in the Karoo Basin of South Africa with his hand in a 255-million-year-old footprint of *Dicynodon lacerticeps*. (*Bottom right*) Artistic rendering of what *Dicynodon lacerticeps* may have looked like in life.

(*Top left*) Janet Voight, curator of zoology, preparing to dive a mile and a half down to the bottom of the Pacific Ocean abyss in the Alvin. (*Top right*) The Alvin, a U.S. Navy-owned Deep Submergence Vehicle operated by Woods Hole Oceanographic Institution (WHOI). (*Bottom*) The deep ocean floor that Janet studies (here captured by the lights of the Alvin) has a thriving but poorly known ecosystem totally devoid of sunlight. We still know less about the bottom of the deep oceans than we know about the surface of the moon.

(*Top left*) Petra Sierwald, curator of entomology. (*Top right*) Shannon Hackett, curator of birds. (*Bottom*) Part of the Pritzker Laboratory for Molecular Systematics and Evolution that is open to public view in the Field Museum's exhibition hall.

Corrie Moreau, curator of entomology and world authority on ants, in Borneo on a collecting trip in 2014. She also started the Women in Science program at the Field Museum, whose mission is to encourage young women to pursue careers in the natural and cultural sciences.

(*Top left*) Gary Feinman, curator of anthropology, discovering a ceramic jaguar dated between 600 and 800 AD in Oaxaca, Mexico. (*Top right*) Gary studying pot shards from Oaxaca. (*Bottom left*) Ryan Williams, curator of archaeology, in Cerro Baúl, southern Peru, where he and colleagues discovered the earliest-known brewery in the Andes region, more than 1,000 years old. (*Bottom right*) Ryan using portable ground-penetrating radar to look for underground archaeological structures in southern Peru.

(*Top left*) Bill Parkinson, curator of archaeology, in a 6,500-year-old archaeological site in Hungary looking for artifacts from when humans were beginning to emerge from the Stone Age. (*Top right*) Bill and colleagues from the Diros Project in southern Greece. *From right to left:* Bill Parkinson, Giorgos Papathanassopoulos, and Anastasia Papathanasiou. (*Bottom*) The discovery by Bill and colleagues in 2013 of an embracing Stone Age couple from 3800 BC.

Bill and his colleagues on the Diros Project focus on the amazing Alepotrypa Cave of southern Greece (shown here), which contains one of the largest Stone Age burial sites ever discovered. Artifacts and over 170 human skeletons have been found there so far, dating between 6000 BC and 3200 BC. The cave is thought to have inspired the Greek legend of Hades and was thought by prehistoric people of the region to be the entrance to the Kingdom of the Dead.

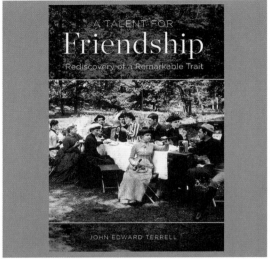

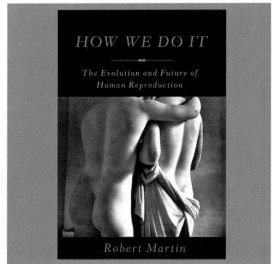

Two museum curators working on the social and physical aspects of human bonding. (*Top*) John Terrell, curator of Pacific anthropology, in his very early years as a curator (in the white T-shirt in a grass hut surrounded by Polynesian natives in 1969) and his 2014 book on how friendship evolved in human society. (*Bottom*) Bob Martin, curator of physical anthropology, and his 2013 book on the evolution of human reproduction and parenting.

Chap Kusimba, curator of African archaeology, sitting among the ruins of Mtwapa, a medieval Kenyan port town that existed from about 1000 to 1750 AD. Chap worked this site from 1986 through 2008 as part of his research on the poorly known cultural history of eastern Africa.

The Field Museum's legendary curator of herpetology Karl Patterson "K-P" Schmidt (1890-1957) with some tiny friends (helmeted iguanas). Photo taken in 1952.

8

The Spirit of K-P Schmidt and the Hazards of Herpetology

As I became more familiar with my contemporary curatorial colleagues in Chicago, I also became increasingly familiar with curators from the museum's distant past. I appreciated their extraordinary dedication and was inspired by some of them for aspects of my own work. One curator in particular, whose name is spoken almost reverently on the ground floor of the museum in the Herpetology Division, is K. P. Schmidt.

Karl Patterson Schmidt (1890–1957)—or "K-P," as he was usually called by colleagues and coworkers—was a curator of amphibians and reptiles (herpetology) at the Field Museum from 1922 to 1955. I can't help feeling some curatorial kinship with K-P Schmidt for historical reasons. We both grew up in the Midwest and both went out east for school. We both spent several years at the American Museum of Natural History before finishing our careers in Chicago. And we both spent over thirty years as curators at the Field Museum. Although I feel this personal connection, I never had the chance to meet him because we were from vastly different generations. He started at the Field Museum twenty-nine years before I was even born. But there is a time-bridging sense of spiritual kinship that comes from shared aspirations and institutional loyalties.

K-P had a distinguished career at the Field Museum, working his way up from assistant curator to chief curator of zoology. When he started in 1922, he had only a bachelor's degree from Cornell University, which he had received in 1916. He was from a bygone era when you could become a curator in a major museum without a PhD. In 1952, just a few years before his retirement from the museum, he received an honorary PhD from Earlham College in Indiana. After he retired in 1955, he remained as a curator emeritus (a distinguished but unpaid position) until his death. His collection of more than 15,000 titles of herpetological literature became the basis for the Karl P. Schmidt Memorial Herpetological Library of the Field Museum. He published many books and scientific articles, named more than 200 new species of reptiles, and was editor of the leading herpetological journal of the day, *Copeia*. In 1956 K-P was elected to the National Academy of Science, one of the highest honors in U.S. science. He was one of the most important herpetologists of the twentieth century.

K-P did a lot of fieldwork in the tropics, which in his day required long trips by boat. He led multiple collecting expeditions to Central and South America, and the herpetology collection of the museum grew from 4,000 specimens to 60,000 specimens while he was curator. Collecting specimens for a zoologist is a little different from that of other curators in natural history museums. Paleontologists use rock hammers, chisels, saws, and shovels; botanists use shears and plant presses; and archaeologists use ground-penetrating radar, shovels, and brooms. For zoologists, the tools of the trade are nets, traps, euthanizing drugs, and shotguns, although many animals are also collected in salvage operations ranging from roadkills to zoo mortalities. Zoologists must take a limited sample of animal species in order to document features that cannot be interpreted on the basis of pictures, video, or observations of living animals in the wild.[1] A new species cannot even be validly named and described without a voucher specimen in a museum to differentiate its distinctness from other species. Some restrictive collecting of particular animals and habitats will always be necessary in order to keep the specimen libraries of the world's natural history museums up-to-date and even to help conserve those species in the wild.. Zoologists in my experience do not enjoy killing things, but their detailed, strategic sampling of biodiversity ultimately benefits the ecosystems that they sample. In that spirit, zoology and botany curators of today continue the legacy of K-P Schmidt and his biological surveys of the world.

In 1923 K-P led a four-and-a-half-month expedition to Central America known as the Captain Marshall Field Expedition, from which he left his field notes in the museum archives. His ship left New Orleans in January and returned to Chicago almost five months later. During this time he collected 1,625 specimens, consisting mostly of snakes, lizards, amphibians, and a rare species of crocodile called Morelet's crocodile (*Crocodylus moreletii*). This species had previously been referred to as a "lost species" because it had only been reported once in 1851 and had not been seen since. Its very existence had come to be doubted in scientific circles. But on his 1923 expedition, K-P rediscovered a population of the species alive and well in the backwater swamps of Belize. Discovering such a legendary animal was a herpetologist's dream. Upon finding the species to be abundant there, K-P became determined to collect a specimen for a biodiversity exhibit in the museum. One evening, working by the headlights of his car shining over the swamp, he spotted a five-foot specimen at the water's surface, visible by the reflection of the car lights in its eyes. Zoologists today have a protocol for how different animal species are most humanely killed for biological surveys, but for some species, a gun is still the only way. K-P crept to the shore and shot at the animal with his .22 gauge long-barreled pistol. He missed, and the crocodile disappeared under the water. Persistence being one of K-P's attributes, he waded out up to his knees in the dark swamp toward the last known position of the animal. As he continued to sludge through the swamp, he suddenly spotted the submerged croc. It was resting on the bottom, no more than two feet away from him. He quickly took aim and fired another shot at it. After traveling through a foot or two of water, the bullet bounced harmlessly off the thick bony head of the crocodile. Even though the shot did no serious damage to the crocodile, it was enough to make the animal severely pissed off. It went into a wild frenzy. As K-P described in his field notes, the animal "came to the surface and dashed madly about in a short figure-eight path, with jaws wide open, and came to a stop just in front of me, still with jaws open." Where this may have scared the hell out of most sane people, or at the very least given them reason to pause, K-P's only fear was the possibility of losing his prize specimen. Instead of backing out of the swamp and away from the crocodile's snapping jaws, he lunged at the animal, grabbing it by the head. K-P writes that the crocodile "was able to twist over and over with astonishing rapidity, necessitating equally rapid changes of hands on his snout to avoid laceration by the sharp projecting teeth." But luck was on K-P's side that day. He finally managed to drag the jaw-snapping crocodile to the shore, where his assistant helped tie the animal up. A plaster mold was eventually made of the prize specimen and a colored cast was made. This cast still stands today in the exhibit hall of the Field Museum, where it has been viewed by tens of millions of visitors through the years.

Although K-P collected all kinds of reptiles and amphibians over the course of his career, his favorite animals of study were lizards and snakes. And it was his interest in snakes that pulled him into a deadly chain of events. It was in September 1957, and Marlin Perkins, director of Chicago's Lincoln Park Zoo, sent a brightly colored African snake to the Field Museum for identification. Perkins (who also hosted the famous *Mutual of Omaha's Wild Kingdom* television series from 1963 to 1985) was himself a good example of the hazards of herpetology. Over the course of his career, he was bitten by three different species of venomous snakes: a rattlesnake, a cottonmouth, and a Gaboon viper. Occasionally Perkins would need help in identifying exotic snakes for the zoo, and the Field Museum was an ideal place to send them, given that it was only a few miles down Lake

Shore Drive and had a world-class team of herpetology experts including K-P Schmidt.

All of the museum's herpetology curators were eager to examine the snake once it arrived at the museum. It was first picked up by curators Robert Inger and Hymen Marx. The snake was a colorful specimen about twenty-six inches in length. Close examination found it to be the venomous boomslang, *Dispholidus typus*. When it was K-P's turn to examine the snake, he grabbed it a little too far behind the head. This allowed the snake enough room to maneuver and bite him on the left thumb. Only one of the snake's small teeth broke the skin, and K-P immediately began vigorously sucking the puncture to remove as much of the venom as possible. Given that it was a very young snake from which only a single three-millimeter-long tooth pierced his skin, and that he sucked much of the venom out of the wound immediately after being bitten, K-P decided not to seek medical aid. Being the curious, opportunistic scientist that he was, he decided to write up the aftereffects of the bite for possible publication. He mentioned to a colleague that "a first-hand report of an untreated bite has special value" since little was known about boomslang bites on humans at the time.

So K-P started a log of his bite-related symptoms. He started out with a description of the snake, saying that based on key anatomical features, it could not easily be identified to species and its identification as a boomslang "was dramatically attested by its behavior." He recorded his symptoms in detail. He wrote in the third person, treating himself as a study subject, with minimal emotional flavor. It soon became apparent that the tiny bite was more serious than he first thought. His handwritten notes (archived in the Field Museum archives) graphically detail his symptoms hour by hour:

> September 25th, 4:30–5:30 PM, strong nausea, but without vomiting, during trip to Homewood on suburban train. . . . 5:30–6:30 PM, strong chills and shaking, followed by a fever of 101.7°. Bleeding of mucus membranes in the mouth began about 6:30 PM, apparently mostly from gums. . . . Urination at 12:20 AM

mostly blood. Mouth had bled steadily as shown by dried blood at both angles of the mouth.

K-P then wrote that he had experienced abdominal pain and nausea all through the night. But by early morning he indicated that his condition was improving. He stated that at around 5 a.m. he was feeling much better and that his temperature was back to 98.2°. He went on to say that he slept comfortably until 6:30 a.m., and at 7 a.m. he enjoyed a substantial breakfast of eggs, toast, and coffee. In fact, he started feeling so much better that at 10 a.m. he called the museum to say he would be back to work the next day.

By early afternoon, things started to deteriorate quickly for K-P. He developed nausea, internal bleeding, and labored breathing. His wife became concerned and called an ambulance. By the time the ambulance arrived, K-P had slipped into a coma. They rushed him to the hospital, but by 3 p.m. on September 26, 1957, the great and fearless K-P Schmidt was pronounced dead from respiratory paralysis, just twenty-eight hours after he was bitten. K-P had taken one too many risks. An autopsy later revealed that he had extensive internal bleeding and massive hemorrhaging throughout his body. A tiny amount of venom from a baby boomslang had caused extensive internal organ damage, surprising even one of the world's top snake specialists of the day with its lethal strength. K-P's detailed and graphic account increased the recorded knowledge about the deadly effects of boomslang venom on humans. We now know that this venom is extraordinarily toxic, even deadlier than that of cobras, kraits, and mambas. It disables the blood-clotting process, resulting in widespread internal bleeding. Because the venom is slow to show serious symptoms, victims sometimes believe (wrongly) that their injury is not serious, until it is too late.

In the end K-P Schmidt, the legendary herpetologist who had published over 150 books and professional papers over his lifetime, contributed to one more paper in death. His sobering notes describing the path to his fatal decline from the snakebite were published by Field Museum curator Clifford H. Pope in the leading scientific journal on herpetology, *Copeia*. This was the same journal that K-P had been editor of for twelve years. On October 3, 1957, some of K-P's notes were also published in the *Chicago Tribune*, where they were referred to as a "Diary of Death." K-P's story demonstrates the scientific curiosity, dedication, and passion that he felt for his profession, even if it contains elements bordering on the macabre. While K-P Schmidt's story certainly does nothing to alleviate my mild phobia of snakes, I will always admire his fearless optimism and dedication to investigating the mysteries of nature. Like the adventures and accomplishments of my present-day colleagues, K-P Schmidt's legendary story impressed and moved me. It is one more inspirational piece in a mix of people and events that contributed to my development as a curator.

The schooner SS *Aurora*. This ship was used by K-P Schmidt in 1923 for the Captain Marshall Field Expedition in Central America, which lasted over four months.

Crocodiles along the shore and in the water photographed by K-P Schmidt during the 1923 Captain Marshall Field Expedition in Belize.

A cast of the toothy 5-foot-long Morelet's crocodile (*Crocodylus moreletii*) that Schmidt wrestled bare-handed to collect for the Field Museum. It is now on exhibit in the Hall of Reptiles at the Field Museum.

K-P Schmidt on the television set of Zenith Radio Corporation in 1940 holding a live snapping turtle by the tail. It was one of a six-episode experiment to see if broadcast television (a technology that was then commercially in its infancy) could be of value as a medium of educational dissemination. Even though K-P never owned a TV in his lifetime, he was one of the pioneers of televised science education.

The boomslang *Dispholidus typus*. (*Top*) A hatchling from Hoedspruit, South Africa; (*Bottom*) an adult male from Mpumalanga, South Africa. This is the species that bit K-P Schmidt in his lab at the Field Museum on September 25, 1957. The bite was a tiny puncture from a baby individual, and K-P thought he had sucked most of the venom out. But twenty-eight hours later he died from the bite, which had eventually caused massive internal organ damage.

September 25.

4:30 – 5:30 strong nausea, but without vomiting, during trip to Homewood on suburban train.

5:30 – 6:30 strong chill & shaking, followed by fever of 101.7°, which did not persist. Bleeding of mucous membranes in the mouth began about 5:30, apparently mostly from gums.

8:30 P.M. Ate 2 pieces milk toast.

9:00 P.M. – 12:20 A.M. Slept well. No blood in urine before going to sleep, but very small amount of urine. Urination at 12:20 A.M. mostly blood, but small in amount. Mouth had bled steadily as shown by dried blood at both angles of mouth.

A good deal of abdominal pain, mainly from gas, continuing to 1:00 P.M., only inadequately relieved by belching. A little fitful sleep until 4 P.M. A.M. when I took an enema (bowels having failed to move the previous day).

Took a glass of water at 4:30 A.M., followed by violent nausea & vomiting, the

A page from the log that K-P started after being bitten by the boomslang on September 25, 1957. By September 26, the venom proved lethal. A *Chicago Tribune* story called this K-P's "Diary of Death." The entire original log is in the Field Museum archive collection.

Fourteen-foot-high sculptures of the Greek Muses built into the four corners of the Field Museum's main hall. They were designed there as architectural monuments to the museum's core missions of scientific research, collections, and dissemination of knowledge. Helping sustain these missions was my primary responsibility as head of collections and research from 2004 to 2013.

9

Executive Management

At some point during their career, curators are often asked to serve in a leadership position, as was I in 2004. After twenty-one years as a curator in the Geology Department, I agreed to become the museum's vice president (eventually senior vice president) and head of Collections and Research (C&R). Together with a small administrative staff, I was responsible for a division that included over 160 paid employees, including all of the museum's curators, the professional staff (collection managers, preparators, conservators, illustrators, lab managers, and technicians), a variety of other specialists (grants officer, repatriation officer, web and technology specialists), and eventually the staff of the library, museum archives, and the Photography Department. In addition, C&R included around 40 resident graduate students from our local university partners, a dozen or so interns, about 250 volunteers, and over 300 people with honorary positions (adjunct curators, emeritus curators, and research associates). It was a diverse group of energetic and dedicated people. Although I missed having as much time for research as I had as a full-time curator, the experience of leading such a diverse group of professionals enriched my perspective on the role of curators, on the mission of natural history museums, and on the enterprise of science as a whole.

Shortly after beginning the job of VP, I came to realize that I had to address three basic levels of management. The first level was my favorite part of the job and in some ways the most complicated: managing the group within the C&R Division. Leading an academic group of brilliant people including curators under the protection of tenure requires a patient and collaborative mode of operation. The challenge is to nurture the intellectual energy of a group of scientists who are talented and know it, without letting things dissolve into complete anarchy. It was sometimes a bit like herding cats. You are not a boss in the traditional sense, as you might be in a corporation or factory. You instead manage with subtle oversight and consultative strategy to facilitate the work of highly skilled people (assuming that you did a good job of hiring the best and most productive people in the first place). You also lead by example, which includes maintaining your own research program at some level. The key is to feed and leverage the passionately charged energy that exceptional scientists innately possess to build dynamic scientific programs of worldwide recognition. In the end, their success is the success of the institution.

I grew to know the museum's broad scientific staff and to see how different operations integrated (or sometimes did not integrate) with one another. Whenever anyone in C&R had a major problem, it usually ended up in my office. Whenever anyone outside of C&R had a problem with someone in C&R, it also ended up in my office. I settled disputes between people and listened to complaints about colleagues, supervisors, departments, C&R, and the museum in general. I listened to emotional pleas for resources, and I received a lot of unsolicited advice on how to do my job better. The experience I had gained working in the complaint department of a Montgomery Ward store during my early years in college came in handy on more than one occasion. You needed to be empathetic, yet you also needed a thick skin. The same drive that enables people to become world-class professional scientists in highly competitive fields can also result in formidable egos, unrelenting determination, and a few rough edges. None of that bothered me very much, because I saw that even the most confrontational personalities were usually driven by a genuine interest in the museum's scientific mission. It is that intense drive that enables research scientists to build great programs and add to the historic reputation of the institution.

The second level of management responsibility I had was to interact with the rest of the museum's executive team to help manage basic museum affairs. The group included the museum president and the museum vice presidents, of which there were many.[1] The most challenging topic we had was to determine the right balance of activity and resources among the divisions of the museum in order to maintain our mission-based goals in a fiscally sustainable way. Unlike a university, the academic arm of the museum (C&R) was not supported by tuition revenue or alumni donations. It is largely dependent on the financial success of the revenue-generating divisions of the institution (mainly the museum's stores, space and event rentals, exhibitions, and fund-raising). C&R needed those divisions of the museum for financial support and public outreach as much as those divisions needed C&R for content and credibility. Part of my job was advocating for the curators, the collections, and the rest of the C&R staff. Although the seven other vice presidents strongly advocated for their own divisions, most of them knew that C&R operations contained the core elements necessary for their division's success and accomplishment of the institution's mission.[2] One of the greatest challenges facing museums today is to keep the science and the non-science divisions in collaborative balance. They are all interdependent and critical to the reputation and sustainability of the institution.

The third level of management responsibility for me as the head of Collections and Research was managing collaborations between C&R and a multitude of entities outside the museum, including federal agencies, universities, private foundations, major donors, and other organizations congruent with our mission. I dealt with many different federal agencies in my role as head of Collections and Research. Each year I met with U.S. Fish and Wildlife officials to explain the reason we had federally protected species in the collection. I would provide copies of permits and make collection records open to inspection by federal officials. In the Chicago office for the U.S. Fish and Wildlife Service, officials usually approved of our mission and our diligence to comply with all federal laws regarding material we had acquired over time. But sometimes it could be more challenging no matter how closely we followed federal laws and standards. Members of the public with good intentions but no knowledge of federal wildlife restrictions would occasionally mail us an unsolicited whale tooth, ivory, a Native American headdress with eagle feathers, or some other item that contained elements from a protected species. We would report it immediately to Fish and Wildlife, and on one occasion rather than rewarding us for our honesty and attempted compliance with the law, the agency issued a citation against the museum for having even received it. It could sometimes be a frustrating and complicated relationship, but we always eventually made it work.

The C&R Division also worked with many other federal agencies, including the National Park Service, the Bureau of Land Management, the National Institutes of Health, the National Science Foundation, the Centers for Disease Control and Prevention, the Department of Defense, the Bureau of Alcohol, Tobacco, Firearms and Explosives, the U.S. Drug Enforcement Administration, the U.S. Agency for International Development, the U.S. Department of Fish and Wildlife, and the U.S. State Department. We developed whatever mission-based partnerships we could that came with federal funding, made collecting on federal lands possible, or enabled us to have regulated objects in our collection. The development and maintenance of good relations with federal agencies was critical. It seemed that I had to sign documents for one federal agency or another almost every day. There are, no doubt, running files on me in government offices somewhere because as head of C&R I was the one who had to sign and hold the firearms permit with the state police (the anthropology collection has everything from handguns to cannons) and the one who signed for the controlled substance permit with the federal Drug Enforcement Administration (the botany collection houses everything from coca leaves of the sort used to make cocaine, to *Cannabis* and hallucinogenic mushrooms). Being in charge of guns and drugs must have been an automatic red flag to some of those federal agencies.

When important political figures would come to the museum, I would sometimes have to meet with Secret Service officials to help set up mandatory security checks. The largest and most challenging of these checks was in May 2012 for the NATO Summit in Chicago. Some events for the summit that would be attended by important political figures were to be held at the Field Museum. The city was filled with several groups of angry protesters. Consequently, I had to select members of the museum staff to coordinate with a small army of Secret Service men and bomb-sniffing dogs to accomplish a seemingly impossible task: a security sweep over the entire collection of the museum in a 24-hour period. I had to select people who could rapidly get federal security clearances and then work with the federal agents and dogs. They would have to go through all of the museums shelves and cabinets. It was quite an ordeal. People worked all through the night, and eventually even the dogs became notably exhausted, deciding that they had had enough of sniffing the countless strange objects in tens of thousands of drawers and cabinets. It was a unique experience that graphically demonstrated the enormity of our collection. During the process, I had some interesting conversations with a couple of the dog handlers. They had recently come off of a tour of bomb-search duties in Iraq looking for improvised explosive devices (IEDs) along the roads. The Chicago job must have been like a vacation for them. I asked one of the men who appeared to be in charge what their normal job duties were when they were in the states. He responded, "Well, you know those guys that dive in front of the president to take a bullet? . . . That ain't us." The inspection process continued on through most of the night with close to eighty people, and somehow we accomplished what needed to be done. As federal agencies go, the Secret Service was an easy one to work with.

Another important external collaboration I oversaw for C&R was the one between curators and local universities. The curators have adjunct professor or instructor appointments and teach at three major universities in Chicago: the University of Chicago, the University of Illinois at Chicago, and Northwestern University. The museum benefits because the universities pay a portion of some curatorial salaries and stipends to curators who teach. Graduate students help with the museum collections and participate on curatorial field expeditions. Museum curators also benefit intellectually from having in-house graduate students, the lifeblood of any academic institution. The universities benefit by being able to offer additional specialty courses without the need to hire more full-time professors. Curators act as student advisors and as a low-cost recruitment tool for attracting students to university graduate programs. Universities also get office space at the museum for their students who wish to take up residence near the collections they use for their thesis research. The synergy between the museum and the universities creates programs of international recognition. Two prime examples include the PhD programs in paleontology and in ecology and evolution in the University of Chicago's Committee on Evolutionary Biology (CEB). These PhD programs were ranked number one in the country by *U.S. News and World Report* in 2008. A third of CEB's faculty consisted of Field Museum curators at the time. Another example of synergistic success is the anthropology PhD program at the University of Illinois at Chicago. This program was originally accredited with the stipulation that the Field Museum's anthropology curators would be part of the university's faculty. Without them, the university faculty would have been too small for a viable PhD program.

As head of C&R, I also became involved in a variety of international initiatives that promoted and applied strengths of the C&R Division. Perhaps the most notable of these was the *Encyclopedia of Life* (*EOL*), for which the Field Museum was one of the five cornerstone institutions (the others being the Smithsonian, Harvard University, the Marine Biological Laboratory at Woods Hole, and the Missouri Botanical Garden). The program began in 2007 as a dream project of E. O. Wilson, the famous curator of entomology from Harvard's Museum of Comparative Zoology. It was seeded with an initial grant of $50 million from the MacArthur Foundation, the Sloan Foundation, and to a lesser degree from the cornerstone institutions. The ambitious goal of the *EOL* was to digitally copy, organize, and coordinate all known information about life on Earth and make it freely available through the Internet on an ever-expanding website.[3] Even the general public could participate online by adding images and other content to help expand the website, making the *EOL* another example of the citizen-science approach to data collection. Members of the general public came to appreciate the importance of biodiversity research not only by using the site, but also by contributing to it. Since its beginning, thousands of people have contributed to the website. My role with the *EOL* was as a member of its executive committee from 2010 through 2013.

The executive committee of *EOL* focused on its long-term sustainability, providing governance and decision making at the policy level. Our main goal was to help the project become a self-sustaining resource with long-term global impact. The initiative expanded to an international effort shortly after the first year. By 2012 there were more than fifty countries collaborating on building the organization. To emphasize the international nature of the project, the executive committee met at locations all over the world, including China, the Netherlands, Mexico, and, most memorable for me, Alexandria, Egypt, in 2011.

I had always wanted to visit Egypt because of its ancient cultural history. There was a great deal of political unrest in the country at the time. The decision to meet in Egypt had been made in 2010, but then in February 2011, the country went through a revolution that resulted in over 800 dead and 6,000 injured. The *EOL* group briefly considered changing the meeting, but as October neared it was decided that the meeting in Alexandria would go on as originally planned. My flight was booked for October 9, with a short stop in Cairo. On the day I left for Egypt, violence began building again in Cairo. On October 10, Coptic Christians collided with Egyptian security forces, which resulted in twenty-six deaths and hundreds wounded. But everything was already in motion for the *EOL* meeting in Alexandria.

By the time I arrived in Alexandria, it was night. The city stretches for twenty miles along the Mediterranean coastline. The air and sound strangely reminded me of Manhattan, with an aroma of seawater and car exhaust and the constant sound of traffic and car horns. I was unable to tell whether the tension that I sensed in the street was just part of the normal urban din, or if it was due to recent events of unrest in the country. But over the next few days, there was no trouble. Our hosts took exceptionally good care of us, and we had a very productive meeting. Still, I remember the odd feeling of walking through the streets and seeing military tanks and heavily armed soldiers in front of buildings near my hotel.

Being in Alexandria, a city founded in 331 BC by Alexander the Great, was a humbling experience. As a museum curator, Alexandria has deep significance for me as the site of the largest most important museum library of the ancient world. The *EOL* meeting was held in the current Library of Alexandria (the Bibliotheca Alexandrina), which is but a monument to the original institution that was destroyed almost 2,000 years ago. The original library was an unsurpassed archive of human knowledge in its day. It was thought to have contained as many as half a million papyrus scrolls (the publication medium of the time), most of which were unique. The library was part of a larger research institution called the Musaeum of Alexandria, where famous thinkers of the ancient world studied, including Archimedes (the father of engineering), Euclid (the father of geometry), Hipparchus (the father of trigonometry), Herophilus (the founder of the scientific method), and Hero (the father of mechanics), just to name a few. This Musaeum was also the original source for the word "museum" as used today. Its only major collection was its enormous library of priceless scholarly documents gathered from around the world. You could say it was one of the world's first major research museums. Its institutional mission was to bring together the best scholars of the Hellenistic world to follow and develop their scientific curiosity in a supportive environment, much like the most successful research museums today.

Astounding discoveries were made in the Musaeum. In the fourth century BC, Aristarchus of Samos discovered that it was the sun, not the Earth, that was the center of our planetary system (heliocentric theory). In the third century BC, Eratosthenes deduced that the Earth was round, not flat, and invented the global system of latitude and longitude. Tragically, the Library of Alexandria and the Musaeum were eventually destroyed, burned once by Julius Cesar in 48 BC and then again in its entirety in 391 AD by order of Coptic Pope Theophilus. A little over a century later, western Europe entered an intellectual Dark Ages that would last nearly a millennium. Some scientific discoveries were lost to the bulk of the world for more than a thousand years until their rediscovery or reestablishment during the European Renaissance in the fourteenth to seventeenth centuries. Columbus reestablished that the Earth was round in 1492 (in the face of controversy with flat-earthers who still insisted that it was flat), and in 1543 Copernicus rediscovered heliocentrism. Other earlier discoveries were eventually reaffirmed as well. But we will never know how many scientific breakthroughs and documents from the Musaeum were lost forever. A visit to the present-day Library of Alexandria can't help but evoke some sense of loss in a curator, librarian, or other steward of human patrimony. Such lessons of history, as well as today's resurgence of museum and artifact destruction in the Middle East,[4] highlight one of the long-term challenges to the international museum community. They must steward accurate records of nature and human civilization in perpetuity regardless of changing political, religious, or economic pressures. Collaborative efforts like the EOL represent one way to help do this through digitization of information and placement of it on the Internet.

The Field Museum remained as an active cornerstone partner in the *EOL* effort until 2013. After six years of growth and development, the *EOL* had evolved from being a start-up program to a functional product. Mark Westneat (Field Museum curator of fishes from 1992 to 2013), who had been the tremendously effective director of the Biodiversity Synthesis Center of *EOL* left the museum to be a professor of biology at the University of Chicago. At that point we felt that our continued investment in the project had passed the point of diminishing return, and the Smithsonian had established a permanent federal budget line to help sustain the *EOL* and be its main driving force in North America. So we concluded our formal involvement with *EOL* to move on to other institutional priorities and new initiatives.

In my role as head of C&R, I had a macroscopic view of the great diversity of curatorial skills, accomplishments, and challenges. I increasingly appreciated how the assets of the C&R Division, both physical and intellectual, formed the core of the institution's mission. We focused on collection-based research and education, while taking on opportunistic institutional collaborations such as the *EOL*, various university graduate and internship programs, and grant-funded projects. I came to recognize what was required to fuel and manage this operation and how C&R needed to fit in with the rest of the museum's enterprises. We needed to be flexible enough to take advantage of mission-based opportunities when they arose yet keep an eye toward long-term sustainability. But the road was not easy.

The most challenging issue I had to face while heading C&R was budgeting for the division during the national financial crisis of 2007–8 and the global recession of 2008–12. As these financial crises affected the nation, they also challenged the fiscal stability of museum operations. Revenue was declining from almost every source, including admissions from attendance, store and restaurant sales, investment earnings on the endowment, government support, and philanthropy. A significant bond debt had also grown as the result of major investments in the building, new exhibits, and expanded new programs. The debt added a new major component to the institution's annual expenses: debt service (interest payments on the debt).

Much like the city of Chicago, the state of Illinois, and the federal government itself, the museum found itself over-leveraged. In the words of the museum's chief financial officer Jim Croft, published in *Nebraska Magazine* in 2014, "It was the perfect storm—and the harbinger of what became several years of massive operating deficits." It was a brutal struggle for sustainability, and it took many years to stop the fiscal bleeding.

Even in the best of times, the annual budgeting process for C&R was like solving a puzzle, but cash flow problems added an extra level of difficulty. The budget for C&R each year was roughly $15 to $20 million, which came from a complicated assortment of more than 140 separate funds. Many of these funds and endowments were restricted for specific purposes by the original donor or funding agency. These restrictions would be for things like "purchase of mineral specimens" or "research in the Pacific" or even "conserving archaeological metals in the museum collection." The trick during the years of financial struggle was to make sure that each fund was used only for what it was restricted for, while using it for budget relief wherever possible to save jobs and sustain operations. The annual budgeting process took weeks of coordinated effort between my office, the scientific department chairs, the museum chief financial officer, and others. As head of Collections and Research, I had to broaden the scale of my external fund-raising efforts, moving from proposals for my own research programs to proposals for the general operation of C&R in order to help keep scientific operations going as smoothly as possible. A lot of people depended on me, and I took the responsibility personally.

I brought in millions of dollars of private foundation money to support C&R positions and research activities while I headed the division, including one of the largest endowment gifts the museum had ever received: $7.3 million to create the Robert A. Pritzker Center for Meteoritics and Polar Studies. The Field Museum has the world's largest meteorite collection of any non-federal research institution, but in 2008 there was no curator or dedicated collection manager for meteoritics. These positions had been vacant since Meenakshi

Wadhwa had left two years earlier. There had been a hiring freeze for research positions at the museum, so the only way I could add a curatorial position to C&R was to find an outside source of permanent funding myself. In 2009, acting on a tip from a research associate of the museum, I contacted Colonel (IL) J. N. Pritzker (Retired), president and founder of the Tawani Foundation. I had heard that the colonel had an interest in meteoritics and polar research, so I put together a comprehensive proposal. I presented the proposal to Colonel Pritzker, and for the next several months we went through a phase of negotiations and careful planning. We reached an agreement that resulted in the meteoritics program being funded in perpetuity. It was an exhilarating experience, and the Planetary Studies Foundation later presented me and the colonel with the James A. Lovell Award for our work on creating the Robert A. Pritzker Center.

I also spent time in Washington, DC, with members of the National Science Foundation looking for ways to increase federal support of collection maintenance for C&R, and I encouraged all of the curators to apply for federal grants to help support operations. I had to be a knowledgeable advocate, salesman, and negotiator for C&R, all in one. It was a challenging, often thankless job, because in times of stress and uncertainty, no one is ever entirely content. But it was remarkably satisfying and reenergizing to watch C&R come through some of the most serious fiscal challenges in the institution's history and remain a significant force in the world of natural and cultural science.

When the financial problems of the museum had become blatantly apparent, the museum trustees and management began looking for substantial ways to bring institutional expenses more in line with its income. Even not-for-profit institutions have to make enough money to pay their bills. An in-depth strategic planning study began taking a hard look at long-term budget issues. We engaged one of the world's most prestigious management consulting firms, McKinsey and Company, to do a thorough review. All "non-revenue-generating" departments (as they were sometimes referred to by the finance department in those days) were on the table. Because curatorial research was one of the so-called non-revenue generators, even it was at risk. One could not sustain curatorial research programs at the Field Museum without first sustaining the museum itself. It was a time of great institutional tension and uncertainty.

After an exhaustive six-month study, it was determined that Collections and Research at the Field Museum constituted essential elements of the museum's brand and therefore were ultimately a necessary core of its future success. It was also found that although maintenance of the collections was a substantial part of the museum's operating expenses, the net cost of curatorial research by the four scientific departments of the museum was near zero because of designated endowment support, federal grants, and other considerations. Research was a museum activity that donors, federal agencies, and private foundations liked to support—and had supported for decades. McKinsey recommended that the museum maintain its dual mission, with one side as the "Science and Collections Institution" and the other side as the "Public Museum." They also recommended restructuring and reorganization within the museum to bring expenses back in line with revenue.

In January 2013, major organizational restructuring began, including within the Science Division. As dictated by institutional policy, I would have to chair an exigency committee. This committee's charge was to determine whether or not it would be necessary to fire tenured curators in order for the institution to return to fiscal sustainability. The breaking of tenure is an extremely serious and risky issue for any academic institution, both legally and with regard to being able to attract the best scientists for vacant positions in the future. The exigency committee could be called only during times of extreme fiscal concern, and this was the first time in the history of the museum that the committee had ever been convened. The committee included the chairs of the four academic departments, the chair of the Science Advisory Committee (a group that functioned as a curatorial senate), the personnel manager, the chief financial officer, and two museum trustees (trustee chairs of the investment committee and the science committee). There were also two attorneys present. It was a difficult period in the museum's history that was graphically portrayed in the media. Newspaper and magazine articles ran headlines such as "Field Museum Cutting Costs, Losing Scientists," "Chicago's Field Museum Cuts Back on Science," "Dinosaur-Size Debt: Field Museum Borrowed Heavily Before Recession; Now It's Paying the Price," and "Field Museum's Debt Forces Reorganization."

The exigency committee met for weeks, carefully reviewing the detailed financial state of the museum and possible solutions. During this period I entered into discussions with my colleagues and the provost at the University of Chicago about the fiscal challenges we were facing in C&R, and I requested their help with curatorial salaries. Committee discussions and outside negotiations went on for more than two months. Then at the end of the tenth week, the university came back with a package of significant salary support for several of our curators,[5] and the members of the exigency committee arrived at a consensus. Given the earlier findings of McKinsey and Company on the importance of research in the museum, the added support from the University of Chicago, plus an incentivized voluntary retirement plan that was devised for curators over fifty-five years in age, it was decided that it would be unnecessary and unwise to fire tenured curators. By early 2014 the retirements and voluntary departures had taken the number of curatorial positions in the museum down to twenty-one through attrition (from a high of thirty-eight in 2004), but no curator had been fired and the scientific enterprise within the museum was now fiscally sustainable.

Chairing the exigency committee was my last act as senior vice president and head of C&R. Once the committee was finished with its charge, I moved back into a full-time curatorial position. After I had served eight and a half years for what was originally a five-year appointment as head of the Collections and Research Division, incoming museum president Richard Lariviere and the board of trustees appointed me to the unique position of Distinguished Service Curator. Once again I would be able to focus on doing scientific research, fieldwork, teaching, and writing. It was yet another personal adjustment in the evolution of my curatorial career, but not an unwelcome one. The eight and a half years I had served as vice president for Collections and Research was the longest anyone had ever held that post at the Field Museum.

Shortly after I left my administrative position, the C&R Division was split into two units: the Integrative Research Center (including the curators and the research operations) and the Gantz Family Collections Center (including the collections staff). Organizational restructuring was still a work in progress at the time I wrote this book. But however operations are divided, or whatever the subdivisions are called, it will continue to be specimen-based research that gives major natural history museums like the Field their deepest scientific credibility.

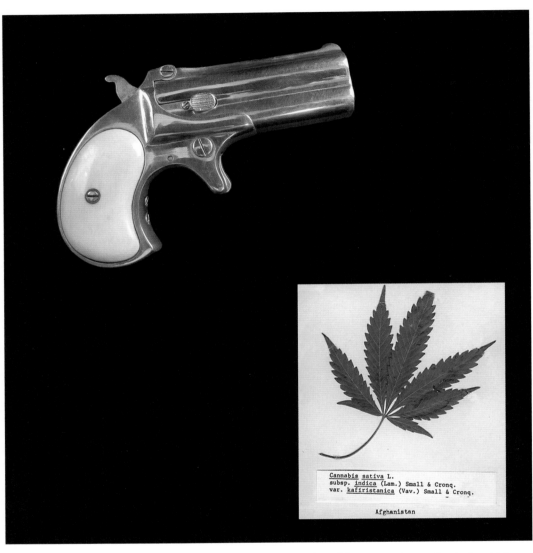

As head of Collections and Research, I was the de facto curator of firearms and narcotic plants, which involved a lot of permit interactions with federal agencies. Shown here are a gold-plated Derringer from the anthropology weapons collection and a *Cannabis* leaf from the botany herbarium.

U.S. Secret Service agents with bomb-sniffing dogs on the roof of the Field Museum in 2012. This was part of a massive security inspection of the museum including our entire collection of 27 million specimens in preparation for the NATO Conference. Here some of us are pausing to enjoy the Chicago skyline before spending the entire night working down through the museum.

Lieutenant Colonel J. N. Pritzker (Retired) and I each received the James A. Lovell Award from the Planetary Studies Foundation in 2009 for our work in conceiving and establishing the Robert A. Pritzker Center for Meteoritics and Polar Studies at the Field Museum. Looking for new ways to fund the operations of Collections and Research was often a large part of my job while heading the division, especially during challenging fiscal years. I found many generous people who were willing to support curatorial-driven research programs.

One of the first visitors to the Robert A. Pritzker Center for Meteoritics and Polar Studies was Bill Anders, astronaut from the 1968 Apollo 8 mission. Bill and his two crew members were the first humans to travel in a spaceship from Earth, orbit the moon, and safely return. Here in the meteorite collection from left to right are me, Bill Anders, and Jim Holstein (collections manager) in April 2010.

There was tension throughout much of Egypt in October 2011 when I arrived for the *Encyclopedia of Life* executive meeting. I could see military vehicles such as this one in the city near our hotel, but our hosts took exceptionally good care of us. Meetings in Egypt, China, and other countries helped establish the *EOL* as an internationally collaborative program.

As head of the Collections and Research division, I occasionally gave tours to dignitaries while they were visiting Chicago. Here, Field Museum Director of Government Affairs Deborah Bekken and I are giving a tour to First Minister of Scotland, Alex Salmond, in 2012.

Mermaid with Goldfish by the famous nineteenth-century stained-glass artist Louis Comfort Tiffany (1848-1933). Over six feet in height, it hangs as a centerpiece in the Grainger Hall of Gems.

10

Exhibition and the Grainger Hall of Gems

In late 2006 Field Museum president John McCarter asked if I would be the curator for an exhibit that was distinctly non-paleontological: a new Grainger Hall of Gems scheduled to open in 2009. I gladly accepted the challenge. Although principally a fish specialist with a PhD in evolutionary biology, I also had a BS and an MS in geology. This made me the closest thing to a mineralogist that the museum had at the time. (Versatility is a useful thing in the curatorial world.) Also, I had been the administrative head of the Collections and Research Division of the museum for almost three years at that point. The opportunity for a creative outlet such as this was an energizing diversion from the day-to-day pressures of managing the C&R Division of the museum. This would be the ultimate interdisciplinary project: a gem hall done with the influence of an evolutionary biologist.

Although major natural history museums today have independent departments of exhibitions that do not contain curators (as mentioned in the preface), curators opportunistically fill the role of content specialist for museum exhibits. They work closely with the museum's exhibition staff to bring authority and cutting-edge science to new exhibits. Exhibitions, sometimes referred to as "edutainment," are one special way that curators can communicate to the general public about why an institution of natural and cultural history is interesting and worthy of their support. Exhibits also bring paying visitors into the museum, and admission revenue is a major source of support for scientific operations.

Since its opening in 1894, the Field Museum has almost always had a gem exhibit.[1] It has long been an important attraction for paying visitors, so my interest in this exhibit had many facets (so to speak). For museums, gems rank right up there near mummies and dinosaurs in terms of exhibit popularity. The last renovation of the gem hall had taken place in 1985, and by 2006 it had started to look dated and dull. There were many world-class gems in the collection, some of which were part of the museum's founding collection from 1893. However, many of even the most prominent of gems sat in darkly lit corners of exhibit cases and provided little of their potential impact.

Being the newly designated curator of gems admitted me to parts of the museum's collection that I had never seen before. For potential display pieces, I was given access to the Geology Department's gem vaults and rare mineral collections as well as the Anthropology Department's high-security gold room and vast archaeological collections. There were world-class gemstones, cut gems, and jewelry pieces ranging from necklaces more than 5,000 years old to pieces created in the twenty-first century. I was also given a list of potential donors to approach for additional resources and jewelry. I would be working with the museum's exhibitions department, and we would have a little over two years to design, build, and open the new hall to the public.

My plan for the new Grainger Hall of Gems was a serious change from its previous renditions. It involved a more exclusive selection of gems, a greater number of natural crystals linking the gems to natural history, and a more systematic organization. Much like the taxonomic classifications I used for animals in my research on fishes, I made a taxonomic classification of gemstones to organize the exhibit, using special shared mineralogical characteristics. In the process, I visually connected the popular, commercially oriented practice of gemology and its gem varieties with the more academic science of mineralogy and its mineral species (yes, mineralogists have species, too). The exhibit case for the mineralogical species *Beryl*, for example, would include all of its gem varieties such as emerald, bixbite, aquamarine, heliodor, morganite, and goshenite.

Although I had a clear systematic organization in mind for the exhibit, my overriding priority would be to focus on the beauty of the material. For each gem variety, I would attempt to bridge the aesthetics of finished gems (human artistry) to the aesthetics of raw crystals (natural history). In addition to reorganizing the exhibit, I would upgrade the type of material in the exhibit so it would focus more on the true gem varieties. I would remove many of the lesser materials that had been included in previous renditions of the hall (e.g., wood, sandstone, feathers, glass replicas, fluorite, and calcite). We had some of the world's most important gems, so why not eliminate the distracting filler material? I would have to plug a few gaps in the museum collection through new acquisitions, and now I had a budget to do that thanks to some new donors. Filling those gaps would require me to travel around the country to various gem and mineral shows, most notably the annual Tucson Gem and Mineral Show. I would also seek out top jewelry designers around the country who might be interested in donating design efforts and gems, and setting some of our world-class gems into innovative pieces of jewelry for the exhibit. We would also find a lighting designer who had learned from the country's best jewelers on how to properly light gems for maximum effect. This was the vision I began with, and I had the strength of the Field Museum's talented Exhibitions Department to help me facilitate it.

We would insure that all new gems added to the exhibit were ethically mined and were not from countries known for trafficking in conflict gems (also called blood gems). In the twenty-first century, there were regions of the world that financed terrorist activity through the sale of gems, most often diamonds. In 2003 the Kimberley Process Certification Scheme (KPCS) group was set up through the United Nations to require member nations not to traffic in conflict gems and provide warranties or certifications that gems were conflict-free. We would work only with jewelers who used gems certified as conflict-free by the KPCS. With that in mind, it was time to begin my search for the pieces necessary to complete the exhibit.

On January 30, 2007, I left Chicago for the Tucson Gem and Mineral Show with the head of Field Museum's Public Programs and the head of the Exhibitions Department. The Tucson Gem and Mineral Show is the world's largest commercial show for natural history items ranging from gems and mineral specimens to fossils. It has been an annual event since 1955. During a few weeks in late January and early February, this show takes over downtown Tucson with approximately 4,500 vendors simultaneously located at more than 45 different hotels, parking lots, tents, and convention centers. Mineral and fossil dealers come from more than 20 different countries around the world to buy and sell items. The show attracts more than 50,000 visitors. Over a billion dollars' worth of gems, minerals, and fossils changes hands here each year. If we couldn't find what we needed, it probably wasn't legally available anywhere. After two days and several hundred vendor booths, we were starting to burn out. We realized that even in a week we could not possibly have inspected the wares of every vendor at this show, but we had already purchased a number of beautiful pieces and were ready to return to Chicago.

One of my favorite acquisitions for the Hall of Gems was a superb grouping of gem-quality aquamarine crystals set in a matrix of white feldspar crystals from Pakistan. It was a specimen preserved just as it had come out of the mine, with fine blue transparent crystals of the finest quality and of the sort used to make fine faceted gems for jewelry (page 248). Such pieces were extremely rare because the gem cutters usually break the transparent faceting-grade crystals off to process into gems for jewelry. Consequently, and somewhat ironically, faceting-grade natural crystals of many gemstones at the show were often priced much higher than the finished gems would have been. I was glad that the gem cutters had missed this specimen, because the natural beauty of the piece could not possibly have been improved by cutting it into finished gems. There was something of a profound peaceful feeling looking into the aqua-blue crystals against the snow-white matrix. In the Grainger Hall, it would be a companion piece to several faceted gems including one of the museum's 148.5-carat deep blue aquamarine gems that Tiffany and Company had set into a pin they designed for the Grainger Hall of Gems (page 249).

Other pieces we acquired in Tucson for the exhibit included a monster gold nugget from Australia weighing four pounds and a specimen of natural star-ruby crystals on a white marble matrix from Vietnam that would add spectacular impact to the new Grainger Hall. Our trip was immensely successful! It had been cash-and-carry transactions with most of the vendors because many of them were not from the United States. Once we had all of the pieces together in the hotel rooms, we began to contemplate how we would get them safely back to Chicago. We knew there was no way we would trust the pieces to ship as checked baggage on the plane. So we decided to discreetly carry it all on the plane with us. The airport security guards of most airports in the United States would be taken aback by a four-pound gold nugget being carried onto a plane in an empty camera case. But in Tucson during the run of the show, the security guard didn't even blink. When the dense fist-size nugget showed up on the X-ray scanning device, he simply pointed to it and asked, "What's that?" We quietly whispered to him, "A four-pound gold nugget." He just responded with "OK."

To further inspire myself about the topic of gems and gemstones, I needed something more. I needed to feel what it was like to be a gem miner. So in June 2008, I set off for the Oceanview mine near Pala, California. It was a working gem mine where they found a variety of highly valued gemstones including tourmaline, morganite, and aquamarine. I made arrangements to prospect for gemstones and to gain firsthand experience of bringing these gemological wonders to the light of day. The Field Museum Exhibitions Department sent one of their developers, Allison Augustyn, along with me. Our host in California was Bill Larson, president of Pala International, who helped us get into the Oceanview gemstone mine.

We were excited to go prospecting. Most visitors to the mine are only allowed to go through the dump piles of picked-over clay outside the mine (for which they pay a fee). But Bill had arranged for Allison and me to enter the mine itself and excavate material from a newly discovered pocket with great potential. The entrance to the mine was about 7 feet high by 7 feet wide and went into a large hill topped with dense vegetation. The cave went hundreds of feet into the hillside, and there was a large yellow plastic duct installed along the roof to pump outside air into the mine to keep the people inside from suffocating. It was not a place for the claustrophobic. The newly discovered pocket of crystal-rich clay was about 100 feet inside of the mine's entrance. Pockets such as this one were the original spaces in which many types of crystals had grown millions of years ago. Later the cavities containing the crystals filled up with clay deposited by groundwater percolating through the mountainside. Today such clay pockets can contain rare and beautiful gemstones.

The air in the cave felt musty and stale, and there was total darkness except where the electric light penetrated. There was also an eerie silence in the cave. You could hear the conversations and breathing of other people in the mine and the sound of their tools working the cave deposits, but the normal background noises of traffic, birds, wind, and airplanes overhead were completely muffled out by 100 feet of overlying rock and dirt. In the cave we slowly filled our five-gallon plastic bucket with the damp clay from the pocket, gently digging it out of the stone crevice with small spatula-like knives, spoons, and hand trawls so not to break any crystals that might be contained within. I struck a large two-pound transparent quartz crystal and was careful to work around it so not to break its pristine surface. It took Allison and me about fifteen minutes to fill a large bucket, and then we carried it out of the mine. Just outside of the cave, we spread its contents over the large table-like screens and then began spraying the whole mess with a high-pressure stream of water from a hose fitted with a hand-held garden sprayer. The clay slowly dissolved in the flowing water and washed away, leaving rocks and crystals of various types on the screen. Other than the large quartz crystal we had found in the cave, the first bucket contained only broken pieces of quartz, chips of white feldspar, and a few miscellaneous chunks of opaque rock. Not very impressive.

We went back into the cave and tried again. We filled up the bucket once more, brought it out of the mine, and spread it on the screen. This time after spraying and dissolving the clay, we found a small smoky quartz crystal, a large optical clear quartz crystal, and a beautiful tourmaline remaining in the screen. There is a certain primeval exhilaration in "hitting pay dirt." Suddenly I could understand what must keep the miners going in these dark caves every day. Over the years in this very mine, there had been single gemstone specimens found that were worth tens of thousands of dollars, and you could never predict when the next spectacular gem would show itself. What a rush! It had been a great trip. I felt newly energized to return to Chicago and continue working on the exhibit.

Once we were back at the Field Museum, we began the final push. We spent long hours sketching case layouts, producing label copy, and optimizing aesthetics with educational content. Then we turned things over to the installation crews, who built environmentally controlled cases in the hall, set up high-security systems, and crafted hundreds of small custom mounts and spotlights for the gems and gemstones. The process took several months working late into the evening. On October 23, 2009, the new Grainger Hall of Gems opened and was an immediate success.

My favorite pieces of antique jewelry in the exhibit was a piece of Edwardian artistry made of platinum and small diamonds featuring a beautifully delicate face carved out of a large blue sapphire. The finished gem weighs 60.2 carats. This not only captivated my newly developed aesthetic appreciation for classic jewelry; it fascinated the geologist in me. Sapphire is a gem variety of the mineral species *Corundum*, which is extremely hard. There are few substances other than diamond that are hard enough to carve a sapphire, and it would have been a long and difficult procedure to carve such a stone with such delicate precision and artistry. I marveled at the patient, specialized artistry that would have been involved to give the piece its delicate, almost angelic quality (page 253).

Then there is the Aztec "Sun-god Opal," a famous piece with gemological beauty and ancient antiquity. It features a 35-carat precious white opal carved with the face of the Aztec sun god. The stone was mined in central Mexico by the Aztecs in the sixteenth century. The Aztecs were enamored with opal, and they called it *quetzalitzlipyollitli*, meaning "stone of the bird of paradise." The "Sun-god Opal" was probably a special ceremonial piece, given the detail of carving. The stone later made its way to the Middle East where it was placed in a gold setting and remained in a Persian temple for several centuries. After changing hands a few more times, it eventually made its way to the Field Museum in 1893.

The oldest piece of jewelry in the Grainger Hall of Gems is the ancient Kish necklace. Kish was an ancient city of Mesopotamia in an area that is now Iraq. The city is over 5,000 years old and is credited for development of one of the world's earliest writing systems, sophisticated mathematics and astronomy, the sail, and the wheel. The Kish necklace is dated at about 2500 BC and is a beautiful combination of carnelian, amethyst, agate, and lapis beads. Carnelian and lapis had deep metaphysical significance to the ancient Mesopotamians, and they valued lapis more highly than gold. The Kish necklace is one of several jewelry pieces that incorporate great antiquity into the exhibit (page 257).

Although the Hall of Gems is primarily an exhibit of gems and gemstones, I also included one precious metal in the exhibit: gold. Gold is the most commonly used precious metal for setting fine gems and has been used in making jewelry for more than 6,000 years. It is ideal for this purpose because its beauty does not tarnish or corrode, and it is malleable enough to be easily shaped and engraved. It has long represented a form of material wealth. The Hall of Gems would be an ultra-high-security room, so I knew that gold objects could be safely exhibited there. There is something about gold that captures that same element of excitement that gems do in many people. Some of the gold came from the Geology Department metal collections (gold nuggets and rare natural gold crystals), some came from the Anthropology Department's gold collection room (ancient gold artifacts from Asia, Europe, and the Americas), and some came from donors whom I convinced to contribute needed items in return for acknowledgment in the exhibit (gold coins, contemporary jewelry). Then there was that four-pound gold nugget we had bought at the Tucson show. It would all be a fitting accompaniment to the gems and gemstones.

Among the most iconic objects that went into the hall was the "Agusan Gold Image" (page 256). This priceless statue is one of the most spectacular discoveries in Philippine archaeological history. It is a solid four-pound, 21-karat gold statue of the Hindu-Malayan goddess Tara made in the thirteenth century. It was found stuck in the mud on the bank of the Wawa River in Agusan, Mindanao, in 1917, where it had been buried for many years. A heavy rain had washed it out of the ground after centuries of being buried and forgotten. Discovered by a woman of the indigenous Manobo tribe, it found its way into the hands of the Agusan deputy governor, who used it to pay off a debt. It seemed to be headed for the melting pot after attempts to have the government buy it for the Philippine National Museum failed. The bullion value of the gold was too much for a country seriously in need of resources. Fortunately, in 1922 the piece was saved. It was purchased for the Field Museum in Chicago by a group including Leonard Wood, the American governor-general of the Philippines; Shailer Mathews, the dean of the Divinity School of the University of Chicago; and Fay-Cooper Cole, Field Museum curator of anthropology. Its existence today demonstrates the importance of maintaining an international perspective on the stewardship of human patrimony. Since coming to the Field Museum, the "Agusan Gold Image" had been periodically featured in several exhibits, viewed by millions of people, and studied by dozens of scientists, but by 2007 it was sitting in storage. It did not seem right to me for such an important piece to be hidden away, so I included it as one of the featured gold objects in the Hall of Gems. Once again, the priceless relic is seen and appreciated by over a million visitors per year.

The renovated Grainger Hall of Gems could never have been done without the generous support of many donors. Part of a curator's job is to help cultivate donors who have an interest in what the curator and the museum are doing. In order to achieve the full vision of what I had in mind for the exhibit, I found people who donated certain pieces of jewelry and top jewelry designers who set some of the museum's world-class gems into pieces of eye-catching, artistic jewelry. I had one thing to offer potential donors in return: long-term acknowledgment of their names in the exhibit. I was fortunate enough to find many generous people who were elated to contribute to the project. For more than a year and a half, I went into every major jewelry store I passed when shopping and asked for donations of jewelry for the exhibit. This would sometimes drive friends and family crazy, but I was driven and fairly optimistic about my powers of persuasion. I was also successful. It was rare event when a jewelry store owner refused to give me what I asked for after I explained that the exhibit would be up for a decade or more and the museum would credit them for their donation in the exhibit. Half a dozen small jewelers donated pieces as a result of this calculated scheme of begging. Coin dealer Harlan J. Berk of Chicago also generously donated tens of thousands of dollars' worth of gold coins that I needed for the exhibit to complement the gold jewelry, gold nuggets, and gold crystals we already had in the museum's collection.

The most remarkable donor of jewelry for the exhibit was Mrs. Thuy Ngo Nguyen. Thuy came to the United States with her husband and infant son in 1975, fleeing the war in Vietnam. She and her husband arrived with few resources, but through hard work and persistence, she and her family built a very prosperous life here. In fact, they were so successful that she started to collect fine jewelry as a hobby. Then some years ago, feeling grateful for her family's good fortune, she began generously donating to various charities in Vietnam and the United States, and cultural institutions including the Field Museum. She donated millions of dollars' worth of jewelry to the Grainger Hall of Gems. It is a remarkable story. Every few months during the period we were planning and renovating the exhibit, Thuy would visit me and the museum's chief financial officer, Jim Croft, wearing jewelry with stones I had told her I was looking for. We would have a nice conversation and sometimes even lunch. Then, as she was about to leave, she would remove all of the jewelry she was wearing and say, "This is for your exhibit." I came to greatly look forward to her visits. Thuy's name is mentioned twenty-six times in the Grainger Hall of Gems, once for each piece of her donated jewelry in the exhibit. The most frequent question I get today from visitors to the Hall of Gems is "Who is Thuy Nguyen?"

Other critically important donors to the exhibit included professional jewelry designers. These talented artists, from New York to Chicago to Oregon, contributed their services to create original pieces of jewelry featuring some of the museum's major gemstones. One group in particular, the Lester Lampert team from Chicago, designed and set twenty-seven pieces for the exhibit. Lester is a uniquely talented fellow who is hard not to like from the first time you meet him. He creates jewelry for the rich and famous: movie stars, famous musicians, and even Pope John Paul. He is also generous. When I first asked him if he would design some settings and set some of the stones for the exhibit, he asked me if he could do *all* of the pieces for the exhibit (and his offer was to do them pro bono). I had to tell him that Tiffany and Co., Ellie Thompson, and Marc Scherer of Chicago, Mish Tworkowski of New York, and several other jewelry designers across the country had already volunteered to do pieces. He then agreed to do as many pieces as I would let him do, and he provided much of the gold and many of the supporting diamonds needed to finish the pieces. One of my favorite pieces that he did for the museum was a ring of 18-karat gold that he set with one of our prize gems: a flawless faceted 27-carat green tourmaline from Maine. He added four carats of small diamonds to the ring to highlight the central gem, and he named the finished piece "Caviar." Lester took great pride in his work, and he named every one of the twenty-seven pieces that he and his team of designers created for the exhibit. ("Caviar" is shown on page 251, and "Mosaic" is on page 247.)

The Hall of Gems exhibit was built primarily to dazzle, educate, and promote the museum's gem and gemstone collection. But public exhibits also need to contain elements of fun. And to that end we added one additional case to the exhibit in the wall near the exit. It is a small case measuring 10 inches high and 24 inches deep that when not in use is covered and fairly invisible to the public. When it is in use, we uncover it to reveal a small, specially lit display case for a single engagement ring. We make this case available by special request for wedding proposals. A hopeful suitor can take his unsuspecting girlfriend through the exhibit, and at the end of it he can surprise her with his bid for nuptial bliss. This idea has proved to be very popular. In the five years since the exhibit opened, there have been 110 proposals made in the Hall of Gems. By all reports, 109 of those were successful on the spot, and one prospective bride-to-be waited until the next day to say yes after giving it some thought. The museum will even sell you an engagement ring if you desire (although if you buy our ring, it is not guaranteed to be a formula for success).

Even though the hall is one of the museum's smallest permanent exhibit spaces, since opening it has attracted almost a million visitors each year. Since opening, the Grainger Hall of Gems has brought in many more new donations of impressive jewels from donors (including my friend Thuy) who hope that their adornments might one day be added to the exhibit. I even published a book featuring the Grainger Hall called *Gems and Gemstones* with my co-developer Allison Augustyn, and it won the 2009 PROSE award for the best book on Earth Sciences. The Grainger Hall of Gems project provided a creative outlet for me on many levels, and it was largely the creative element that attracted me to museum work in the first place.

Digging for tourmalines and other gemstones in the Oceanview mine near Pala, California. (*Top*) Exhibit developer Allison Augustyn and me at the entrance to the mine. (*Bottom*) One of the mine's employees and me about 100 feet inside of the narrow mine shaft, removing buckets of mineral-rich clay from pockets within the wall.

 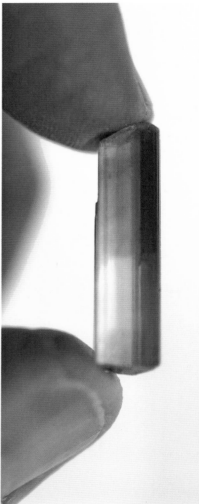

(*Left*) Once we removed the mineral-rich clay from the mine, we spread it on wire screens and sprayed it with water to dissolve away the clay and expose the crystals and pebbles contained within. (*Right*) The sort of crystals one can find if they are lucky enough include this natural gem-quality, tricolor tourmaline.

(*Top*) Faceted gems made from a tricolor tourmaline crystal like the one on the previous page. (*Bottom*) The above gems set in rose gold with small diamonds, by Lester Lampert's jewelry design group for the Grainger Hall of Gems. They named this piece "Mosaic."

Natural aquamarine crystals with a white feldspar matrix as mined from a cave in Pakistan. It's hard to beat the beauty of nature. This specimen is about eight inches wide and is on exhibit in the Grainger Hall of Gems. It was one of many pieces that I used in the exhibit to tie gems and jewelry to natural history.

Faceted 148.5-carat aquamarine set in a platinum with gold accents and white diamonds called "The Schlumberger Bow." Based on a design by Jean Schlumberger, Tiffany and Company fashioned this pin especially for the Grainger Hall of Gems using one of the Field Museum's premier-cut aquamarines.

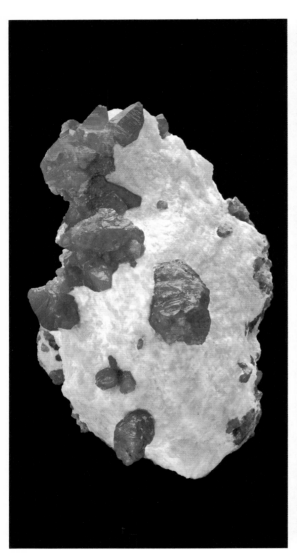 

Asian star ruby specimens on exhibit in the Grainger Hall of Gems. (*Left*) Natural crystals that formed within a white marble matrix. Specimen is from Vietnam and measures 5 inches high. (*Right*) A finished cabochon of about 8.5 carats.

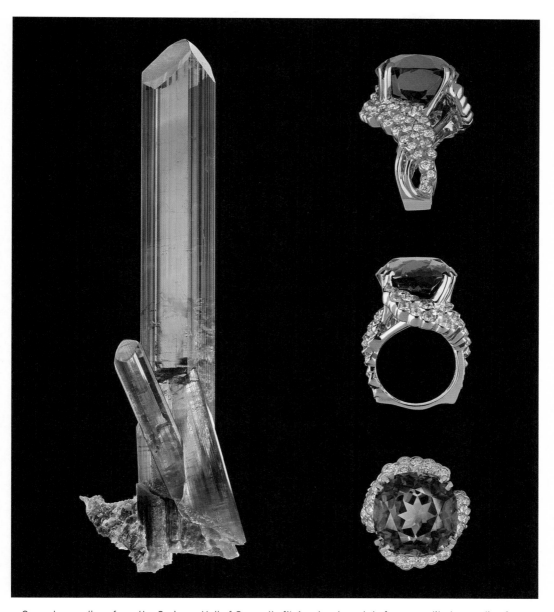

Green tourmalines from the Grainger Hall of Gems. (*Left*) A natural crystal of gem-quality tourmaline from Brazil about 3 inches in height; (*right*) three views of a 27-carat faceted tourmaline gem set into a ring designed by Chicago's Lester Lampert that he named "Caviar." Lester designed and set more than twenty-five of the museum's world-class gems for the exhibit.

A natural blue sapphire crystal weighing 263 carats and measuring 6 centimeters in height. From Kashmir, India. One of many uncut gemstone crystals I put into the Field Museum's Grainger Hall of Gems.

A stunningly beautiful 60.2-carat blue sapphire carved into the form of a delicate face, set in an Edwardian 18-carat platinum pendant, accented with white diamonds. From the early twentieth century by Cartier. On exhibit in the Grainger Hall of Gems.

Where paleontology meets gemology: a 55-carat amber pendant with gold chain. Amber is fossilized tree resin, and the amber in this piece is about 30 million years old with several 30-million-year-old insects fossilized within it. Mined in Mexico and made into jewelry in the early twenty-first century. On exhibit in the Grainger Hall of Gems.

The Aztec "Sun-god Opal." A 35-carat opal that was mined in Mexico by the Aztecs in the sixteenth century. (*Top view*) The opalescent play of color is visible from the front. (*Bottom view*) Shown at an oblique angle to highlight the surface that is carved with the image of the Aztec sun-god. This was one of several pieces that I used in the exhibit to tie gems and jewelry to cultural history.

The "Agusan Gold Image," a priceless thirteenth-century statue of a Hindu-Malayan goddess made of 4 pounds of solid 21-karat gold. Statue height about 7 inches. This piece is one of the most important archaeological artifacts ever discovered in the Philippines. On exhibit in the Grainger Hall of Gems.

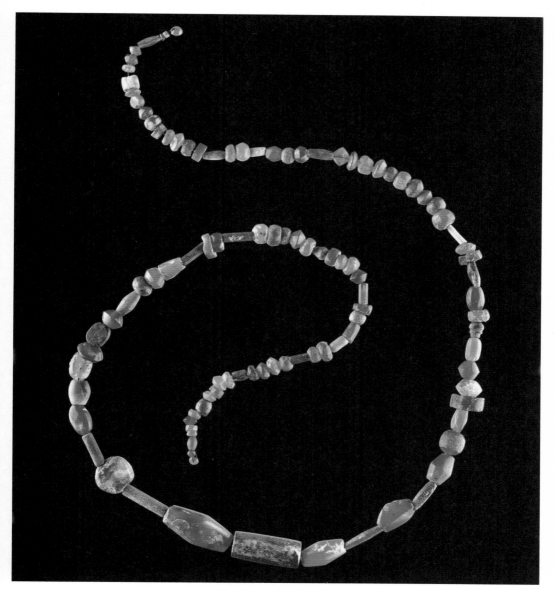
Ancient necklace of amethyst, carnelian, lapis, chalcedony, and agate from the Kish archaeological site in Iraq dated at about 2500 BC. On exhibit in the Grainger Hall of Gems.

Franz Boas posing for a sculptor in 1894 during construction of a diorama on the Kwakiutl Hamatsa cannibal ritual. Boas, known today as "the father of American anthropology," was one of the Field Museum's first curators. He started the museum's human remains collection.

11

Grave Concerns

One of Chicago's largest mausoleums is in the Field Museum. The museum's anthropology collection includes the remains of more than 6,000 human beings, ranging from a 15,000-year-old fossilized human skeleton from southern France to skeletons of executed prisoners from the twentieth century. It includes an iconic collection of Egyptian mummies, Native American skeletons over 3,000 years old, a shrunken head collection from Ecuador, and even a few 50,000-year-old Neanderthal fragments. The human remains collection is a rare and irreplaceable resource that instills a sense of reverence in those who care for it. While serving as head of collections and research, I came to appreciate its scientific value as well as its ethical challenges.

There are important reasons for a major natural history museum to have such a collection. The remains are critical for studies of cultural and physical anthropology. After all, it is hard to know who we are as a species without first knowing who we were. The human remains are also important for a large array of educational training programs ranging from human anatomy to forensics. More recently, human remains are serving as an important resource for DNA analysis and tracking the evolution of disease in order to develop better defenses against it. But because these are *human* remains, there is an extra degree of sensitivity surrounding them and special respect is paid to how they are managed. These specimens are the remains of someone's ancestors, and to many cultures they have great spiritual significance. But the ethical standards we have today took many years to develop.

In the nineteenth and early twentieth centuries, the collecting of human remains, particularly those of indigenous people, was a competitive effort among major museums of the world, as well as by private pot hunters.[1] Many specimen procurement and collection management practices of museums in those days would be unacceptable today. Museums in North America rarely collect human remains anymore. On the rare occasions today that they do (mostly as salvage operations or housing orphaned collections discarded from other museums), they do so under highly regulated conditions. Modern museum policy, federal regulations, and evolving ethical standards have all but stopped the collecting of human remains in North American museums, making existing human remains collections in museums all the more important as a scientific resource. Collections of this type could never be duplicated.

Two past curators were of particular importance to the study and ethical management of the human remains collection at the Field Museum. One was the museum's first anthropology curator, Franz Boas (1858–1942), and the other was one of my colleagues, Jonathan Haas (1949–). Boas helped establish the first human remains collection of the Field Museum, and his work on it did much to combat race prejudice in the late nineteenth and early twentieth centuries. Nearly a century later, Haas was part of a team that helped establish local, national, and international policies on the repatriation of human remains. These policies provide for the return of culturally sensitive remains that were inappropriately collected in the past to

Native American tribes, native Hawaiian organizations, and kinsmen who wish to see them reburied.

Franz Uri Boas was born in Minden, Prussia, in 1858, and in 1877 he began his university training at Heidelberg University. Boas had a scrappy personality that did not suffer insult easily, and his Jewish background occasionally drew challenges from a culture of growing anti-Semitism in Germany. During Boas's time at the university, men fought duels with swords to defend their honor with the objective being to cut your opponent's face. Boas was a highly skilled swordsman, but his face bore the scars of many duels, and became part of the Boasian mystique to his graduate students at Columbia University. Most photographs of him show the right side of his face, which was protected by a mask during duels. Photographs of the left side of his face have usually been retouched to hide the old wounds. Boas was described by one of his students, A. L. Kroeber, as "a rugged, massive, powerful individual of great caliber who drove his engine through to the accomplishment of whatever the task in hand seemed to be." The combative personality that had helped him survive as a student in Germany was often expressed later in life as hostility to authority. His strong convictions gave him a reputation of being difficult to work with. But his convictions were also his strength, and he became the founder of anthropology as an established academic discipline in the United States.

Boas was an immersive field anthropologist who is often credited for establishing the importance of field-based anthropology in a discipline that had previously been dominated by armchair speculation. One region where Boas spent much time doing fieldwork was on Baffin Island with the Inuit people (some of whom call themselves Eskimos). The German academic system that Boas had come from argued that the Inuit people of Baffin Island were at a much more primitive stage of cultural evolution than the idealized German people. Boas strongly disagreed. He argued that the Inuit culture and the western European cultures were each advanced in their own way; they were simply different by necessity in order to best survive in their respective environments. Within the context of the northern Arctic environment, he reasoned, the Inuit people were more advanced than western Europeans. His ideas were not immediately accepted.

Boas immigrated to the United States from Germany in 1887 to become assistant editor for the journal *Science*. He had felt increasingly alienated by anti-Semitism in Germany. He came to the United States with doctorates in both psychology and physics, as well as postdoctoral work in geography. In the United States he continued to challenge the popular ideologies of scientific racism that existed at the time. Boas pushed to reorganize cultural anthropology into more of an environmental and geographical context (race variation as a product of environmental and social learning factors) rather than the popular racial evolution model of the time (implicitly inferior non-European racial types evolving up the evolutionary ladder to an idealized western European race). His views challenged many major museums, zoos, and other institutions that attracted visitors by sensationalizing exotic racial types as primitive savages. Although it seems incredible today, in the late nineteenth and early twentieth centuries, the popularity of human zoos still prevailed, particularly in Europe but also to a limited extent in the United States. A disturbing example of this was in 1906 when the New York Zoological Gardens displayed a young Congolese "pygmy" taken from central Africa after being purchased from slavers for a pound of salt and a bolt of cloth. For weeks he was in a cage in the zoo's monkey house, drawing thousands of New Yorkers and commanding news headlines such as "Bushman, One of a Race That Scientists Do Not Rate High in the Human Scale."[2] This was just the sort of racial chauvinism that Boas could not abide by, but it would take many years for this institutional practice and social mind-set to fade.

In 1889 Boas was appointed as the head of a newly created Department of Anthropology at Clark University in Massachusetts, where he conducted anatomical research on human skulls in order to disprove the assumed deterministic connection between biology and behavior. Eventually he clashed with the university president, claiming infringement of his academic freedom. He resigned in protest in 1892. In September 1893 when the forerunner of the Field Museum was founded as the Columbian Museum of Chicago, Boas became its temporary curator of anthropology and technically the museum's first curator of anthropology. Along with Boas came his collection of 400 human remains, forming the base of the museum's physical anthropology collection. At this point the museum had not yet opened to the public, and its exhibits were still in the construction phase. Boas spoke out with increasing vigor against the idea that human behavior and aptitude could be understood through an evolution of racially categorized anatomical features. Anthropologists of that day defined a systematic classification of human physical features recognizing more than a hundred types of race, almost as though they were each a distinct species.[3] The ideas of an evolutionary ladder of races and the idealized European persisted. Cranial size of the skull was one of the major features thought to support this theory. But the research of Boas, which was based on hundreds of human skeletons and on skull measurements of over 17,000 people, contradicted this. He showed that head shape and size was relatively plastic in nature and varied due to changing environmental factors such as nutrition and health. He was an early civil rights advocate. He gave public lectures about how his research proved that the African race was not innately inferior to the European race. He contended that the scientific racism of the day was extrapolated from a popular misreading of Darwin suggesting that humans were descended from chimpanzees. Boas pointed out that Darwin never said that humans evolved from chimpanzees but instead that humans and chimpanzees were equally evolved, sharing a common ancestor somewhere back in their evolutionary histories.

Boas wanted his position of curator at the museum in Chicago to be made permanent, but the founding leaders of the institution were not impressed with his vision of anthropology or his ideas about exhibit organization. Through the early months of 1894, Boas continued to have personality clashes with museum administrators as well as the influential president of the nearby University of Chicago, William Rainey Harper, who complained that Boas did not "take direction" well. In April of the same year, about two months before the museum officially opened to the public, the thirty-five-year-old Boas was replaced by William Henry Holmes as curator of anthropology. Holmes had previously been a curator at the Smithsonian, and in 1894 he was a more prominent scientific figure than Boas. Holmes also lacked the political baggage of Boas, who by this time had fallen out of favor with several of the trustees and administrators of the museum. No one at the time anticipated that Boas would eventually far eclipse Holmes in the field of anthropology.

In 1895 Boas became a curator of anthropology at the American Museum of Natural History in New York. About the same time he also became a professor at Columbia University, where he started to build a base of students and anthropology programs. At the American Museum, Boas once again hoped to reorganize anthropology exhibits within an environmental context rather than an evolutionary one, with tribal groupings rather than races. Once again a clash of differing philosophies on matters ranging from the layouts of museum exhibits to perceived infringements on academic freedom brought him into conflict with administrators of the institution. The friction grew until May 1905, when he resigned from the American Museum, this time never to work for a museum again. It is another example of a groundbreaking curator challenging the status quo, like Colin Patterson and Gary Nelson discussed in chapter 1. Boas's ideas were ahead of their time and much more consistent with those of today's anthropologists and systematic biologists.

Having resigned from the American Museum, Boas was now able to concentrate his energies on his professorship at Columbia University, where in 1899 he started the first anthropology PhD program in North America. His doctoral students included another famous curator from the American Museum, Margaret Mead, as well as several other highly influential anthropologists. His work and legacy of students transformed the study of anthropology into an established academic discipline, and today he is known as "the father of American anthropology." By 1926 Boas's students headed every major university Department of Anthropology in the United States. His stand against scientific racism also led him to speak out publicly on social issues. He was a strong advocate for racial equality already in the late nineteenth and early twentieth centuries. He argued boldly in speeches across the country that his research clearly falsified the assumptions of innate racial inferiority. It was culture, not nature, which explained the differences among the peoples of the world. In his 1963 book *Race*, Thomas Gossett wrote: "It is possible that Boas did more to combat race prejudice than any other person in history."

The study of human remains today has evolved well beyond looking for anatomical differences among races. The human skeleton is filled with information about human interactions with their social and physical environments. This includes clues about what earlier societies ate, how hard they worked, whether they were subjected to violence, and some of the sicknesses they contracted. The ancient history of human remains holds clues to finding treatments for modern diseases. DNA can be extracted from the bones to identify genes that may have made ancient people resistant to diseases such as TB, syphilis, malaria, arthritis, influenza, and others. Professors also use these collections to teach the finer points of forensic work to students as well as personnel from the FBI, military, and police departments. And much can be learned about cultural history by the study of burial practices and funeral objects associated with mummies and other archaeological skeletons.

As the types of research done with human remains have changed, so have the methods and practices of collecting human remains. In the nineteenth and early twentieth centuries, major museums that had anthropology departments were racing to add human remains (mostly skeletons and mummies) and funerary artifacts to their collections. There were few collecting restrictions, and methods of collecting lacked the ethical standards of museums today. There was sometimes a fine line between scientific collection of human remains and simple grave-robbing. Human skeletons and particularly human skulls were freely bought, sold, and traded. There was an assumption that the scientific end justified the means. Even Boas was ethically enigmatic in this regard. Although he was leading the fight against scientific racism and also a prominent advocate for racial equality, he freely bought, sold, and traded human remains, and even supported himself occasionally by doing so. He also dug up many a graveyard skeleton in his early days to use for his research, and over 430 human remains in his possession constituted part of the Field Museum's founding collection in 1893. The rush to build anthropological collections was partly due to competition among the major museums of the time for exhibit material. There was also a fear by these museums that exotic indigenous cultures and artifacts were disappearing faster than they could be documented as part of human history. William Holmes, the first chief curator of anthropology at the Field Museum, felt an urgency to build the collections while the opportunity still existed. He was quoted saying that the Field Museum could not afford to take a subordinate place to the other museums around the world. During his time at the Field Museum, the anthropology collections grew at an enormous rate. Before he resigned from the museum in 1896 to return to the Smithsonian, he passed his sense of urgency down to his curatorial successor, George Dorsey.

Dorsey was an even more aggressive collector than Holmes. He was ambitious, demanding, and the greatest museum builder of the period. He was a relentless collector himself, and he pushed his subordinates relentlessly. He sent his assistant curators all over the world with a mission to aggressively collect archaeological material for the museum. He expected his assistant curators (four of whom had been trained by Boas) to endure hardships and sacrifice comfort in order to acquire extensive collections of antiquity for the museum. It was Indiana Jones on Steroids. In a letter dated January 31, 1900, from Dorsey to assistant curator Stephen Simms, Dorsey advised:

> When you go into an indian's house and you do not find the
> old man at home and there is something you want, you can do one

of three things: go hunt up the old man and keep hunting until you find him; give the old woman such a price for it as she may ask for it running the risk that the old man will be offended; or steal it. I tried all three plans and I have no choice to recommend.

Many months later Dorsey became unhappy with Simms's rate of productivity, and in another letter to him Dorsey wrote:

> Remember that you are after stuff. . . . You are absolutely compelled to get to out of the way places, to suffer inconveniences and on occasions suffer hardship.

Dorsey clearly saw himself in a competitive race against time. In a letter to a private donor named Stanley McCormick about collecting antiquities in Arizona, he wrote:

> It is none too soon that we are taking up this work, for it is my firm belief that within three or four years there will not be a ruin on the Hopi Reservation that has not been ransacked and devastated by eastern institutions or by Arizona relic hunters.

Dorsey was driven. In 1898 he was arrested in Washington State on a charge of grave robbing. Working for Dorsey could be dangerous. In 1907 he sent an assistant curator William Jones on a lengthy expedition to the Philippines to collect information and artifacts from the Ilongot people. The Ilongot were a tribe that still practiced headhunting. After nearly two years of immersing himself in Ilongot society and collecting anthropological specimens for Dorsey, Jones was murdered by three angry tribesmen. Not to waste any collecting efforts, Dorsey immediately sent assistant curator Simms to the Philippines to recover what was left of Jones's collection and field notes. Jones was left buried in the Philippines with a simple grave marker.

The federal Antiquities Act of 1906 slowed down the mass pillaging of graves on federal land for a while by private collectors (referred to as pot hunters), but the victory was short-lived. In the

1970s two cases against pot hunters were dismissed by judges ruling that the terms of the federal Antiquities Act were vague and unenforceable. Finally in 1979 a new, less ambiguous statute was passed called the Archaeological Resources Protection Act.

Native Americans were concerned about more than continued looting. They were also concerned about the remains of their ancestors and funerary objects that were already in museums. Many Native North Americans and Native Hawaiians believe that the remains of their ancestors and associated spirits do not belong with the living (e.g., as museum exhibits or collection items) and that such a desecration disturbs the spiritual well-being of the deceased individuals. They therefore wanted remains and grave objects that had been removed in the past to be returned to their spiritual repose in order to return the deceased ancestors to peace. Throughout the 1970s and 1980s, Native American groups and museums passionately debated, collided, and sometimes collaborated in search of a solution. That solution finally came in the form of the Native American Graves Protection and Repatriation Act (NAGPRA), which passed into federal law in 1990. This law requires federally funded agencies and institutions to return Native American human remains and associated objects to appropriate descendants who request them for reinterment. It also makes it a criminal offense to traffic in Native American human remains.

In 1989 Jonathan Haas was hired as vice president of the Collections and Research Division and curator of anthropology at the Field Museum. He had received a PhD in anthropology from Columbia University in 1979, the same PhD program started by Boas eighty years earlier. Jonathan worked closely with many Native American tribes for his research, and he was part of the group including Native Americans, archaeologists, and museum officials that crafted NAGPRA. He had a vested interest in that his research program sometimes involved human remains. One of his past research projects involved using skeletons to study human warfare and the relative frequency of warfare through history. The question he wanted to answer was "Is warfare an integral, inescapable part of human culture?" Archaeologists working for the Pentagon were advising the U.S. Department of Defense that warfare was inherent to the human species, and because it was based on biology, it was unavoidable. Jonathan's research based on human remains and archaeological cave art came to the conclusion that warfare was not an inherent component of humanity, but rather a cultural phenomenon that comes and goes as material circumstances change. Jonathan Haas and colleagues surveyed over 2,500 known human skeletons from 400 different archaeological sites around the world that were older than 10,000 years. He was looking for signs of violent death such as skull fractures from a blow to the head, forearm fractures from blocking such a blow, or stone points imbedded in bone. This sample of early human remains included most of the well-preserved material that was known from that time range. Of the entire sample, only four sites showed any signs of violence. In a subsequent review of materials *younger* than 10,000 years, they commonly found evidence of repeated warfare in archaeological sites around the world (much like Bill Parkinson's group found in Alepotrypa Cave, as discussed in chapter 7). This suggested to Haas and his coauthor that there was some fundamental change in society that occurred around 10,000 years ago, resulting in humans becoming more warlike. They concluded that if we could better understand why people changed, perhaps we could find new ways of avoiding war in the future. It is an interesting hypothesis and a line of research that is still debated today. The hypothesis is not far from Margaret Mead's famous theory that human warfare is an invention, not a biological necessity.

In 2002 Jonathan Haas hired the Field Museum's first full-time repatriation specialist, Helen Robbins,[4] who is now also an adjunct curator in anthropology. Helen is one of the most empathic scientists I have ever met, and she puts her heart, as well as her academic strengths, into every repatriation project that the museum undertakes. She forges effective connections and communications between the museum and Native American groups, building bridges of understanding through what is often a difficult process. She also works effectively with anthropology curators, who provide necessary consultation on cultural sensitivities, collection significance, and validity of repatriation requests. The Field has since repatriated many remains, some of which are covered by NAGPRA, and some of which are not. The museum has done its best to set a positive example for other museums around the country.

The return of human remains to the Inuit people of northern Labrador, Canada, in 2011 was one repatriation action in which I was involved as senior vice president for Collections and Research. In 1927 the remains of twenty-two Inuit individuals had been inappropriately dug up by an inexperienced archaeologist from the Field Museum named William Duncan Strong. Strong had been hired right out of graduate school to, among other things, collect human remains. During a fifteen-month museum expedition in northern Labrador, he and the expedition captain, Donald MacMillan, were made aware of an Inuit graveyard in an abandoned Moravian mission called Zoar. The mission had been abandoned in 1894 as being unsuitable for habitation. Strong and his assistant were directed by supervisors and the leader of the expedition to collect the remains from the graveyard. Strong was an untenured assistant curator and felt compelled to comply. He and his assistant dug up twenty-two marked graves, after which he wrote, "Neither of us ever had such a nasty job before." Then things became even darker. After hearing complaints of grave robbing from nearby locals, the police magistrate ordered Strong to return the remains to their original burial plots. Strong appeared to comply by refilling the graves, but he secretly took the bones back to Chicago, where they would remain for the next eighty-three years. Details of the incident were eventually buried in Strong's field notes in the deep archives of the Smithsonian, and the event became forgotten within the institution for decades.

More than half a century later, Strong's notes were discovered by an Inuit working with an anthropologist at the Smithsonian who later brought the incident to the attention of Inuit officials. Eventually the incident came to the attention of Field Museum repatriation director, Helen Robbins. The Field Museum's repatriation team then began several years of collaborative discussions with representatives of the Labrador Inuit (the Nunatsiavut Government). The team's goal was to figure out the best and most respectful way to return the ancestral remains and to help rectify unethical actions of days long past. We worked with the Inuit community to decide where and when the remains should be reburied. We made it clear that this unethical behavior was the product of a time long since gone, and that this type of behavior would be inconceivable at the Field Museum today. Over the months a friendly respect was built between the museum and members of the Inuit community. Although we could not change what had happened in the past, we moved forward to build a positive relationship with each other. We even attended a hockey game together after finding out the Nunatsiavut representatives were huge fans of the Chicago Blackhawks hockey team. Fortuitously, Blackhawks owner Rocky Wirtz was a member of the Field Museum's board of trustees, and he provided prime seats for a game to the visiting Nunatsiavut representatives and the museum staff working with them.

The remains were carefully placed in burial boxes that the museum had specially made. The Inuit representatives said prayers over the remains in a brief ceremony held in the museum. Then the burial boxes were then transported to the airport, where the Field Museum had arranged air transportation to Nain, Labrador. Along with the remains, there was an official letter of apology from Field Museum president John McCarter to the Inuit people of Labrador, saying:

> We are deeply saddened by this incident. While Field Museum employees of today did not commit this wrong, we recognize that these actions did not comply with ethical and archaeological practices, either past or current.

In response, a Labrador Inuit official said:

> While we can't change the past, we can do the right thing now and ensure that these individuals are returned to their rightful resting place.

In May 2011 the remains arrived in Nain, Labrador. From Nain they were transported a few weeks later on a long-liner fishing boat to the abandoned mission site of Zoar, about a four-hour trip. There, on a hill near the original Inuit graveyard across the isthmus from the mission site, a large grave was dug and the remains were ceremoniously reburied on June 22, 2011. Helen Robbins and Joe Brennan (museum general counsel) officially represented the museum in the ceremony, along with more than eighty Inuit and several Innu people, and two Canadian Mounties. It was an emotional, solemn affair. The sky was gray and cloudy to begin with, but during the ceremony the sun cut through the clouds. Then an amazing thing happened that was observed by many there including Field Museum attendees. As an eagle flew overhead, two black bears came out of the woods at the original mission site. Both bears remained still and appeared to be listening and watching. This was extremely symbolic to many people who were there. Black bears are thought to harbor the spirits of deceased Innu ancestors. It projected a powerful feeling to the participants that the souls of these people would now be at rest.

Although Zoar remains abandoned today, a newly built white picket fence now surrounds the burial site, and a stone tablet was placed on the grounds recording the names of people buried there and the events leading up to their reinternment. The carver of the tablet is the famous Inuit artist John Terriak, whose direct ancestor was among those reburied. At the end of the ceremony, the Nunatsiavut Government presented a letter to the Field Museum representatives signed by the Nunatsiavut president and its Minister of Culture, Recreation, and Tourism. The last few lines of this document read:

> Throughout the entire repatriation process, the Nunatsiavut Government and the Field Museum have developed a good relationship based on mutual respect. It is our hope that we will continue to build on that relationship.

> On behalf of all Labrador Inuit, the Nunatsiavut Government graciously accepts the apology from the Field Museum.
> We forgive you.

This long and careful process took nearly five years from start to finish, and the end result was the sort we strive for in any repatriation effort. Today the Field Museum has maintained this connection with the Labrador Inuit people. We are developing ways to collaborate with the Labrador Inuit community to develop a mutually beneficial exhibit, research, and educational opportunities.

Some human remains are still in a controversial gray area when it comes to how a museum should steward them, such as the shrunken head collection of the Field Museum. We have eleven shrunken heads made by the Jivaro people (or Shuar, as they self-identify) from the Amazon region of Ecuador and Peru. The Shuar call the heads *tsantsa*, and they believe that they have magical power, or *tsarutama*. They believed that taking the head of their enemy and making it into a *tsantsa* harnesses the spirit or soul of the slain enemy and prevents it from taking revenge on its murderer. Shrunken heads were made in the upper Amazon region during the late nineteenth and early twentieth centuries as part of an elaborate ritual following battle. Sometimes shrunken heads were also made of revered members of their community as a way to keep their spirits with the living. The practice reflects on the complex intertwining of life and death in the Shuar belief system. In the early twentieth century after many shrunken heads had been obtained by museums as well as private collectors around the world, an international market developed. The heads were freely bought, sold, and exchanged. As demand outstripped supply, head-hunting practices actually increased in the upper Amazon region to fill the demand! All of this came to a close after the Peruvian and Ecuadorian governments outlawed the traffic in human heads. Counterfeit shrunken heads then began to enter the market. Imitations were made from corpses dug up from grave sites and morgues. Others were taken from monkeys or other animals and made to look human. Many fakes are so realistic that it has been estimated that more than half of all shrunken heads in museums today are not genuine. Today the Shuar still produce replica heads from goat skin that they sell to tourists. Two authentic shrunken heads were on exhibit in the Field Museum for many years, and they became somewhat iconic exhibit pieces of the institution in the late twentieth century. In fact, it is the exhibit I remember most from my first visit to the museum as a teenager. The shrunken heads even made an impression on a recent president of the United States. In his book *Dreams from My Father*, Barack Obama writes:

> At the Field Museum, I saw two shrunken heads that were kept on display. They were wrinkled but well preserved, each the size of my palm, their eyes and mouths sewn shut, just as I would have expected. They appeared to be of European extraction: The man had a small goatee, like a conquistador; the female had flowing red hair. I stared at them for a long time (until my mother pulled me away), feeling—with the morbid glee of a young boy—as if I had stumbled upon some sort of cosmic joke.

We took the heads off of exhibit in 1997, as the museum continued to grow more focused on developing better ethical standards for the human remains collections. There was no legal reason to take them off exhibit. There were no formal letters of protest from the Shuar people. But the exhibit context had focused on the heads as though they were mere curios meant to frighten. The anthropology curators of the museum thought the exhibit was not an accurate representation of the Shuar belief system and was culturally misleading. The shrunken heads were an integral part of religious rituals related to warfare, and the anthropology curators felt that displaying them without proper explanation was disrespectful. Taking the exhibit down generated some backlash from the media, who accused us of going overboard in the interest of being politically correct. However, we determined that we would need a more considerate presentation if we were to put these back on public display. The heads represent a complex issue of competing, mission-based ethical responsibilities for the museum.

Not all human remains in the collection are as culturally sensitive as those of Native Americans or as crassly presented as the shrunken heads once were. One of the most socially and ethically acceptable exhibitions of human remains today in museums are the Egyptian mummies. Egyptian and many South American cultures have little problem with the exhibit of ancient ancestral remains in accredited public museums. This is fortunate, because there is little doubt that along with dinosaurs, mummies are the main lifeblood of exhibit operations in many natural history museums. At the Field Museum, the most popular traveling exhibit that we ever hosted in our over 120-year history was the King Tut exhibit on loan from Egypt, featuring the world's most famous mummy: Tutankhamun (c. 1341 BC–c. 1323 BC). This exhibit came to Chicago twice (although Tut himself was not in the second show), and each time people lined up for city blocks to get in to see it. The museum was filled to capacity on many days during its run. The Field Museum also has an extensive mummy collection of its own. During the 1893 World's Fair, the Field Museum acquired a large collection of mummies from Egypt, Peru, Ecuador, and Chile ranging from 800 to 5,500 years old. Many of them have not been on display since 1893 because they are so fragile.

In 2011 curator of anthropology Robert Martin, museum conservator JP Brown, and adjunct curator Jim Phillips led a project to make computed tomography (CT) scans of several of the museum's mummies that had never been unwrapped. A medical facility called Genesis Medical Imaging (GMI) in a northwestern suburb of Chicago donated the services of their CT machine to the museum. They loaded the machine into a large, specially adapted semi-trailer and drove it to the west parking lot of the museum, where the work could be done without moving the mummies very far. Then the Field Museum team and technicians from GMI took CT scans of ten human mummies that were far too fragile to physically unwrap. The scanning allowed the scientists to digitally unwrap them to determine the age, sex, and causes of death of the mummified individuals, and also to decipher the techniques that were used to mummify the bodies. It also revealed many burial objects that were previously unknown, such as god and goddess figurines, offering pots, and various food items.

Today the proper curation of human remains collections presents special challenges for natural history museums and involves the coordinated efforts of many people. Conservators and collection managers preserve and document human remains in the collection. Curators, repatriation specialists, and museum legal experts answer ethical and legal questions about which specimens can be made available for study or exhibit, which specimens must be kept hidden from public view for ethical reasons, and which remains must be repatriated to the place of origin in response to legitimate claims. In spite of many challenges, the maintenance of existing human remains collections in major museums is more important than ever. These remains comprise one of the most empirical sources of human history that we know of. They provide directly observable records of human biology, human belief systems, and society's approaches to death. And anything that leads to a better understanding of human history has the potential to help us plan a better future.

Grave marker on Luzon Island in the Philippines for Field Museum assistant curator William Jones. He was murdered there in 1909 by members of the indigenous people he was studying and collecting from, and he was left buried there. Jones was the first Native American to receive a PhD in anthropology and was a student of Franz Boas. Stephen Simms, the assistant curator sent by Dorsey to claim Jones's collections and notes, erected this simple marker over Jones's burial site. Photo from 1910.

Curator of anthropology Jonathan Haas with Native American artifacts. Jonathan was part of the team that helped establish national repatriation policy for North American museums, including NAGPRA.

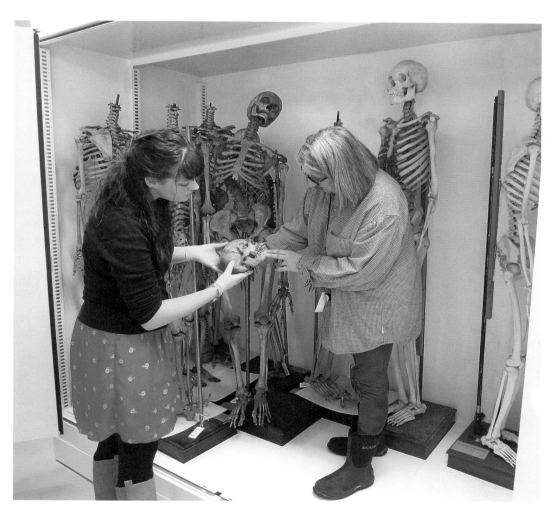

Loyola University anthropologist and Field Museum research associate Anne Grauer, with student Eden Lantz, examining skeletal material from the Field Museum's human remains collection.

Meeting in Chicago with Inuit representatives on May 25, 2011, to organize returning the remains of twenty-two Inuit individuals. Shown here from right to left: Ryan Williams, Field Museum curator and chair of anthropology; Jamie Brake, archaeologist for Nunatsiavut; Johannes Lampe, Nunatsiavut Minister of Culture, Recreation and Tourism; and me representing the museum's Office of Collections and Research.

On June 22, 2011, the Inuit remains arrived in northern Labrador for reburial in one of the Field Museum's most successful repatriation initiatives. (*Top*) Camp set near the abandoned mission of Zoar for the reburial ceremony. (*Bottom*) A white picket fence was built around the burial site as part of the memorial.

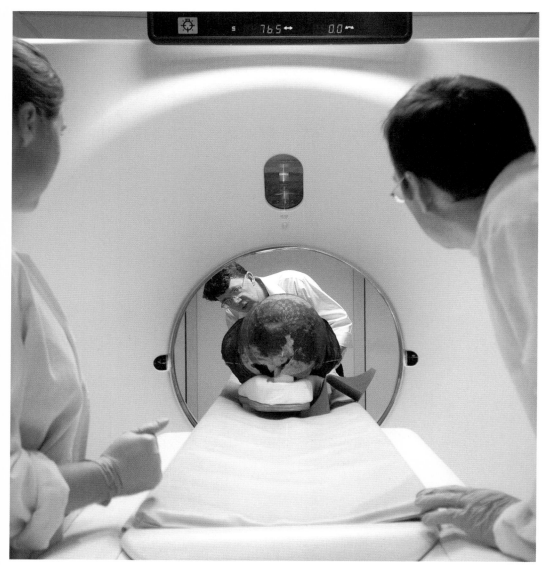

J. P. Brown, anthropology conservator (center over mummy), guides an Egyptian mummy through the CT scanner to waiting technicians on the other side. Image taken in July 2011. This was an ideal way to digitally "unwrap" fragile mummies that were thousands of years old and too delicate to physically unwrap (see next page).

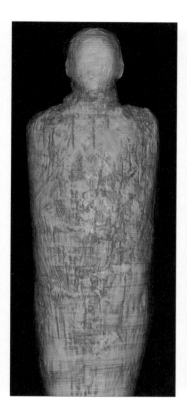  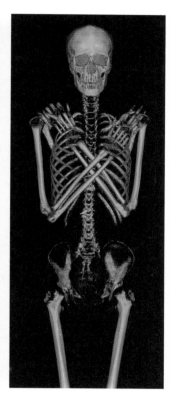

Progressive CT scans of an Egyptian mummy from the Twenty-Sixth Dynasty (664 BC–525 BC). Field Museum specimen too fragile to physically unwrap. Left scan shows the surface wrappings; middle scan is deeper, through the wrappings and showing skin of head and shoulder areas; right scan is the deepest, penetrating through the skin and wrappings to show the skeleton. Mummy was revealed to be a male who died at about seventeen years of age of a chronic illness.

The legendary man-eaters of Tsavo, one of the Field Museum's most popular dioramas since 1926. Originally, they were images used to symbolize human dominance over nature, but gradually they became a symbol of human efforts to conserve such species.

12

Hunting—and Conserving—Lions

Among the museum's more than 20 million zoological specimens, there are two that hooked me during my first visit to the museum as a high school student: the man-eaters of Tsavo. These two Kenyan lions were each nearly ten feet in length and four feet high at the shoulders, and their reputation as ferocious man-eating carnivores is admittedly what first caught my interest as a teenager. They were credited with eating anywhere between 28 and 140 people over a period of only nine months in 1898. The story of the hunt for the man-eaters became one of the most popular tales of the twentieth century. It was the basis for an internationally best-selling book published in 1907 by John Henry Patterson that is still in print today called *The Man-Eaters of Tsavo and Other East African Adventures*. It has also been the basis for four movies, the most successful of which was the 1996 Paramount Pictures film *The Ghost and the Darkness*, starring Michael Douglas and Val Kilmer. There has always been a fascination with animals that can kill and eat an entire person in a matter of hours. And over the course of my career at the museum, I came to appreciate another side of lions: their importance to the African ecosystem.

The man-eaters of Tsavo are connected with three Pattersons of particular significance to the Field Museum, each of whom provides a different but interesting story. The first was John Henry Patterson (1867–1947), the man who killed the lions and later sold their skins to the museum. He was an engineer in addition to being a big-game hunter and a British soldier. He served in the army for thirty-five years (1884–1920), where he rose to the rank of lieutenant colonel. The second Patterson was the colonel's son, Bryan (1909–1979), who came to the Field Museum with the lions at the young age of seventeen and over time rose to become chief curator of the Geology Department. And finally there is Bruce Patterson (1952–), a longtime friend and colleague of mine who has been curator of mammals for more than thirty-five years. Bruce has worked with the original man-eater specimens as well as the modern-day lions in Kenya to promote conservation of the species for the good of the African ecosystem. I found each of the Pattersons to be inspiring to me in their own way. The story of Tsavo lions begins in the late nineteenth century in East Africa, with the British expansion into Africa.

In the early 1890s, the imperial British Railroad set out to construct the Uganda Railway through eastern Africa as a way to strengthen British domination of the region. The expansionists within the British Empire were convinced that a railroad into uncharted African territory was both a road to riches and a way to strengthen its global domination. There was formidable opposition to the endeavor by many in the British Parliament, particularly liberals who pronounced that the government had no right to drive a railway through country owned by the native people of the region, the Waata, Kamba, Taita, Kikuyu, and the Maasai. British tabloids of the day referred to the push into East Africa as the "Lunatic Line, going from nowhere to utterly nowhere." But there was no swaying the forces of British imperialism in the nineteenth century, and the

British government set out to construct a railroad all the way across British East Africa (known today as Kenya, Uganda, and Tanzania).

Colonel John Henry Patterson was commissioned by the Uganda Railway Committee in London to oversee the construction of a bridge over the Tsavo River. The Uganda Railway project was enormous, involving over 30,000 Indian laborers, 3,000 of whom were stationed near the Tsavo River. They camped in tents that collectively stretched for miles. The Kamba origin of the name Tsavo is "place of slaughter," which turned out to be quite prophetic. Shortly after Colonel Patterson's arrival in Tsavo in March 1898, workers on the project began disappearing at an alarming rate. Sometimes assorted bones or other body parts remained. It soon became apparent that two lions were preying heavily on the project campsite, especially after one of the lions tore Patterson's trusted Sikh servant, Ungan Singh, to pieces. Attempts to deter the lions with campfires and heavy thorn barriers were futile. The lions seemed fearless and ready to do almost anything to satisfy their hunger for human flesh. They leaped over or crawled under large thorny barriers, fearlessly taking screaming workers from their tents into the night to eat them. The lions even came into the hospital tent for victims. They were so large and strong that they could carry off a full-grown man. The continued carnage caused hundreds of the remaining workers to flee Tsavo and halted the railroad project. The two lions literally stopped the British Empire in its tracks. At that point, Britain's House of Parliament demanded that the project get moving again. So Colonel Patterson, an experienced tiger hunter from his previous military experience in India, set out to hunt down these voracious predators, vowing to "rid the neighborhood of the brutes." He set traps and spent many nights waiting in a hastily built tree platform trying to ambush the lions. But his mission turned out to be much more difficult and dangerous than he had anticipated.

The dense thorn scrub of the man-eater's territory made moving through the brush nearly impossible for humans, while the agile and elusive lions seemed to have little difficulty. In his 1907 memoir, *The*

Man-Eaters of Tsavo, Patterson wrote that "[the lions'] methods became so uncanny, and their man-stalking so well timed and certain of success, that the workmen firmly believed they were not real animals at all, but devils in lions' shape." The lions would eat their prey at night within earshot of the camp, noisily chewing through human flesh and bone, nearby but unseen in the bush. While he perched in his rickety tree platform late at night hoping to ambush the lions, he could hear the lions creeping around him for hours. The big cats' superior night vision gave them a distinct advantage over humans in the dark of night. At times it was not clear who was hunting whom.

Then, more than eight months after his arrival, Colonel Patterson finally had his first opportunity to stop one of the man-eaters. A Swahili man ran into camp shouting out, "*Simba! Simba!*" (Lion! Lion!). Patterson followed the man back to lion tracks that led into a dense thicket. Returning to camp, the colonel summoned a large group of workmen to bring drums, tin cans, and other things with which to make loud noises. He brought them back to the thicket where he suspected the lions lay waiting and had them encircle all but a place where he would be poised with his rifle. With their noisemakers, the workers flushed one of the lions out of the thicket about fifteen yards in front of Patterson. He raised his rifle, aimed squarely at the growling beast preparing to charge, and to his horror the rifle misfired! Fortunately, the lion was scared off by the sound of the misfire and by the noise from the workmen. As the lion retreated, Patterson fired again, and this time scored a hit. The lion still bounded away, wounded and no doubt angry. The colonel spent the following several nights nervously waiting in ambush for the lion to return, believing that it would eventually come back to finish off a donkey it had been eating days before. After some tense days, the lion eventually wandered into the sights of Patterson's rifle for one last time. This time his bullet found its mark, and the lion fell. The huge feline measured nine feet eight inches in length and took eight men to carry back to camp. Upon his arrival back in camp, hundreds of Indian railroad workers set off in wild jubilation. In *The Man-Eaters of Tsavo*, Patterson recalled the scene at camp:

> They surrounded my eyrie [platform] and, to my amazement, prostrated themselves on the ground, saluting me with cries of "Mabarak, Mabarak," which I believe means blessed or savior.

This was quite a contrast to the tensions that had been building between Patterson and the workers before. Some workers had not taken kindly to Patterson's strict standards of performance when he first arrived. Two plots by workers to murder Patterson had previously been thwarted only by some loyal workers who warned Patterson beforehand.

However, the ordeal with the man-eaters was not yet over. The other lion still remained. It took Patterson another three weeks of concentrated effort to kill the second lion in what was even a more harrowing conclusion. Like the first lion, the second one occasionally stalked Patterson as he stalked it. He and the lion continually changed roles, between being the hunter and the hunted. It was a contest of determination and survival in which only one would likely survive. One morning Patterson spotted the lion dragging a dead goat into the bush. He followed it and put two shots into its shoulder at close range, but the lion bounded off wounded and wasn't seen for ten days. Then on the evening of December 28, Patterson came face-to-face with the angry lion once again. When it charged him, Patterson shot the animal seven more times, stopping it only five yards from him. The second lion measured only two inches shorter than the first. The killing of the second lion stopped the attacks on the railroad workers. Those who had previously fled in fear for their lives returned, and the bridge construction resumed in February 1899. The grateful workers presented Colonel Patterson with a silver bowl with the following inscription:

> Sir,
>
> We, your Overseer, Timekeepers, Mistaris and Workmen, present you with this bowl as a token of our gratitude for your bravery in killing two man-eating lions at great risk to your own

life, thereby saving us from the fate of being devoured by these terrible monsters who nightly broke into our tents and took our fellow-workers from our side. In presenting you with this bowl, we all add our grateful prayers for your long life, happiness and prosperity. We shall ever remain, Sir, your grateful servants,

Baboo PURSHOTAM HURJEE PURMAR,
overseer and clerk of Works, on behalf of your workmen.
January 30, 1899

Colonel Patterson wrote in his best-selling book on the ordeal that he considered the bowl to be the hardest-won, most highly prized trophy of his career.

Nightmares haunted Patterson in the weeks to follow. A short time after both lions had been killed, he was exploring some rocky hills a couple of miles northwest of the bridge (although in his field notes he said southwest, which confused many later scientists trying to retrace his footsteps). Making his way along paths worn by rhinos and hippos over the years, he stumbled on a deep cavern littered with human bones. In *The Man-Eaters of Tsavo*, he writes:

> Round the entrance and inside the cavern I was thunderstruck to find a number of human bones, with here and there a copper bangle such as the natives wear. Beyond all doubt, the Man-eaters['] den! . . . I had no inclination to explore the gloomy depths of the interior, but thinking that there might possibly still be a lioness or cub inside, I fired a shot or two into the cavern through a hole in the roof. Save for a swarm of bats, nothing came out; and after taking a photograph of the cave, I gladly left the horrible spot, thankful that the savage and insatiable brutes which once inhabited it were no longer at large.

After Colonel Patterson hastily left the cave, its location became lost for nearly a century. It was not until 1997 that it was rediscovered by a collection manager from the Field Museum, Tom Gnoske,[1] adjunct curator Julian Kerbis-Peterhans, and Samuel Andanje from the Kenyan Wildlife Service. The cave's location was rediscovered only after they figured out that the orientation of John Patterson's map in his field notes was off by 90 degrees. I visited Tsavo in 2007 while on a trip to explore initiatives with the Kenyan National Museums with fellow curator Chap Kusimba. We traced some of John Patterson's footsteps, including the man-eaters' den. On the day we visited the cave, the Kenyan Wildlife Service sent an armed guard with us for protection against possible lion attack. The cool dark air inside the cave and the sight of the guard's M-16 helped me appreciate what Patterson must have felt as he entered the cave littered with human bones. It was both chilling and exhilarating at the same time.

After Colonel Patterson left Tsavo and wrote his best-selling book *The Man-Eaters of Tsavo*, he toured the world giving lectures on his experiences in Africa. He had the man-eaters' skins and heads made into trophy rugs, as was the custom for big-game hunters of the time. He kept them in his home for many years, until their existence came to the attention of the Field Museum's president Stanley Field. Field expressed a strong interest in obtaining them for the museum. The story of the Tsavo lions continued to spark public interest, and it was thought that the skins could be restored into full-size lion mounts for an exhibit, and the heads that still contained the original skull bones could become useful research specimens. Patterson agreed to sell them to the Field Museum for the sum of $5,000 (nearly $70,000 in today's dollars). The museum taxidermist used the pelts to reconstruct lifelike mounts of the man-eaters in the iconic diorama. The two lion mounts are actually smaller than the animals were in life, due to the skin having been trimmed for use as trophy rugs, but that has never dampened their popularity among museum visitors.

Shortly after the museum acquired the two lions from Colonel Patterson, it also acquired the services of his seventeen-year-old son, Bryan. The colonel had become friends with the director of the Field Museum, and in 1926 he arranged for Bryan, effectively a high school dropout, to be hired by the museum to do whatever work the museum wanted him to do. The colonel seemed to think that the museum would be a good place for his son to learn some honest skills and make something of himself. This humble beginning turned out to be the start of a most remarkable career.

Bryan's first three years at the museum were as a fossil preparator in the Geology Department. He was an insatiable reader and developed a passion for books and articles about paleontology. Occasionally he would take a class at the University of Chicago, but he never enrolled in any formal college degree program. By 1930 he was made a departmental assistant, and by 1937 he was promoted to curator of paleontology. Bryan became a U.S. citizen in 1938, and in 1944 his curatorial career was interrupted by World War II when he enlisted in the army. He was wounded during the invasion of Normandy and captured by the Germans. He escaped from a Nazi prison camp twice but was recaptured both times. Luck was evidently with him, and he survived to be released after the war ended. Bryan returned to his curatorship at the Field Museum, where he continued his long, productive career, and in 1948 he was elected president of the Society of Vertebrate Paleontology. In 1955 he left the Field Museum to accept a tenured professorship in vertebrate paleontology at Harvard University that came with a curatorship at the Museum of Comparative Zoology. He remained a curator and a professor at Harvard for the next twenty years. In 1963 he was elected to the prestigious National Academy of Sciences. The many achievements and accolades of Bryan Patterson were truly remarkable, all the more so because he never formally graduated from high school and had no college degree other than the honorary one that Harvard granted him when they hired him as a tenured professor.

From 1963 through 1967, Bryan led annual expeditions to Kenya, the country of his father's famous exploits with the man-eaters. During his 1965 expedition, he discovered a fossil that brought him international fame—the first bone of a human-like creature that would later be called *Australopithecus anamensis*. At the time this was the oldest known specimen of a hominid (the family including humans and some closely related fossil species). The popular press referred to it as "the oldest ape-man," "the oldest human," or the "Kanapoi hominid." The specimen was only the elbow end of a humerus (arm bone), but it was virtually indistinguishable from that of a modern human. It was particularly interesting because of its old age, first thought to be 2.5 million years old and later found to be 4.1 million years old. In the years to follow, it was cited in the literature by both evolutionary biologists and young Earth creationists. Evolutionists used it to give an impressively early age of origin for human-like creatures. Creationists, mistaking the fossil for *Homo sapiens*, used it as evidence that the dating techniques of scientists couldn't be trusted. Eventually many other bones thought to belong to Bryan's ape-man species were discovered, showing the species to be very different from *Homo sapiens*. The material was described and collectively named *Australopithecus anamensis* by Meave Leakey and colleagues (1995), although Bryan Patterson's fossil from 1965 (first reported in 1967) was the first specimen of this species ever discovered.

The third Patterson of this chapter is my colleague Bruce Patterson, who had no relation to the previous two Pattersons.[2] His research focuses mostly on bats, rodents, and smaller mammals, and also on host-parasite coevolution (how the evolution of a host species correlates with and influences the evolution of their parasites). Bruce's research also includes work on the modern-day lions of Tsavo. His research on the Tsavo lions began in 1998, when he joined fellow Field Museum colleagues Julian Kerbis-Peterhans (adjunct curator), Chapurukha Kusimba (curator of African archaeology and ethnology), and Tom Gnoske (assistant collection manager) in the Tsavo Research Initiative (TRI) to study Tsavo's mysterious lions and their environment. With his TRI collaborators, Bruce focused on questions about the evolutionary relationships of the lions, the absence of manes in Tsavo males, and why the legendary man-eaters of John Henry Patterson's day were so actively preying on humans. Bruce has also worked hard on the environmental conservation issue of helping modern-day Tsavo lions and humans live together more peaceably. In the late twentieth century, the fear *of* lions in Tsavo had begun changing to a fear *for* lions in Tsavo and the possibility of losing them to extinction. Bruce enlisted an approach that I and many other curators use for fieldwork: citizen-science. Bruce did this effectively by partnering with the Earthwatch Institute. This organization underwrites scientific fieldwork by raising funds and enlisting everyday citizens as volunteers to work with the scientists. Between 2002 and 2009, Earthwatch provided Bruce with 542 volunteers from forty different countries to travel to Africa and work with him. All volunteers paid their own way and contributed something extra to support the costs of the expeditions. This enabled Bruce to make three to four trips to Africa each year and drive thousands of miles of roads looking for lions. By leading these groups of volunteers to collect data, he could pursue his research on the Tsavo lions at no additional cost to him or the museum. His citizen-science approach also educates the public on the importance of what he is doing. Leveraging one's curatorial salary and institutional support with fund-raising and enlisting the aid of the public is an important part of a curator's job.

Some of the first questions Bruce and his colleagues tried to answer were questions of relationship and taxonomy. Were the modern-day Tsavo lions a previously unrecognized new species of lion? They look markedly different from other modern-day lions in that they lack a mane, just like the man-eaters. Pleistocene lions are thought to have lacked a mane (based on 30,000-year-old Chauvet cave paintings from southern France). It was suggested that the Tsavo lions could be a different, more ancient species than the other African lions. Bruce spent several trips studying both maneless lions and the full-maned lions of Africa in their native habitats. As part of his research, he tranquilized some of the lions to enable him to fit them with radio tracking collars and to harmlessly take tissue samples for DNA analysis. Once they woke up, he could track them for years with GPS to determine the extent of their normal range. The DNA analysis indicated that the Tsavo lion's genetic makeup did not differ appreciably from nearby lions with manes. After years of studying the lions in their natural habitats and captive lions in zoos, he concluded that the lack of a mane correlated with individuals whose range was within hotter, dryer areas. His colleagues Tom Gnoske and Julian Kerbis-Peterhans reached the same conclusion independently in a separate study done at the same time. The maneless Tsavo lions were *Panthera leo* and not a distinct species from other living lions with manes. Manelessness is simply an acquired variation within modern lions in response to heat and lack of water.

Bruce was also interested in why the famous man-eaters of Tsavo were so predaceous on humans. Past theories had suggested that they may have been accustomed to finding dead humans along the Tsavo River crossing where slave caravans bound for Zanzibar had routinely discarded dead captives. There had also been an outbreak of a severe cattle disease in 1898 that had devastated the lion's usual prey, making humans an opportunistic replacement food source. Bruce and another colleague (a dentist from Waukegan, Illinois) studied the skulls of the man-eaters that the Field Museum had acquired with the skins. They discovered that the first of the man-eaters had severe tooth and jaw problems that would have made it extremely painful to make the killing bite across a struggling animal's throat, especially considering the lion's normal diet of buffalo and wildebeest. Later studies indicated that it was probably the lion with the tooth abscess that had done most of the man-eating. Humans would have been softer and easier prey, with the added plus that they are slow and don't see very well in the dark. And as Bruce says, once the lions tasted human flesh, man-eating very quickly became a routine way of life for them.

Bruce's interest in Tsavo lions goes far beyond their history of carnage. He and his colleagues are focused on social and ecological issues resulting from human competition with lions. The Tsavo lions, as well as all African lions, seem to be headed toward extinction. The rates of decline are alarming, with the number of African lions falling 30 percent in just the past two decades (three lion generations). Ranchers north of the Tsavo region drive their cattle through the Tsavo reserve on their way to market in Mombasa. During the long trek, the Tsavo lions will kill some of the livestock. The herdsmen and ranchers retaliate by killing the lions. Between 2001 and 2006, more than a hundred lions were killed in the Tsavo region by local residents and cattle ranchers.

The extinction of lions within the African ecosystem could prove disastrous. All ecological systems on the planet consist of a synergistic network of interacting species. It includes a food chain ranging from the smallest feeder organisms (e.g., bacteria and algae) to the largest (e.g., humans and lions). Alterations at the low end of the food chain can shock the entire system, causing a ripple effect up the line, but alterations at the upper end can be equally damaging. Lions control populations of herbivores in a balanced equilibrium that was established hundreds of thousands of years ago. Scientific research indicates that removal of top predators from an ecosystem can upset the entire network, disrupting systems that produce food, hold human diseases in abeyance, and even stabilize climate. This phenomenon, sometimes referred to as a top-down force in nature or an ecological chain reaction, has been documented several times. One example was the overhunting of sea otters around the Aleutian Islands in the 1990s that led to a spike in the population of their historic food source: kelp-feeding sea urchins. The increased urchin numbers led to the collapse of the undersea kelp forests, and the loss of the kelp forests resulted in declining numbers of many species dependent on that habitat. The otters were a keystone species to a complex ecological network. Similarly, removal of wolves from Yellowstone caused a chain reaction in the region that ultimately led to the decline of willows, fishes, and amphibians. One theory from the 1960s called the green world hypothesis contends that plants prevail in the world today because predators hold the herbivores in check. The lions play a key role in maintaining stability of the ecosystem in Africa, and ultimately in the world as a whole.

How do we create an optimal balance between the welfare of lions and people? Bruce and his colleagues tallied records of hundreds of attacks on livestock over a period of four years to search for a solution. He concluded that during the dry season, the lions stayed near the water holes and preyed on wild game that gathered to drink. During this time there was plenty of game, and the lions didn't appreciably bother the cattle. But when the rains came, the native prey dispersed and became harder to find. Then the lions turned to more opportunistic prey: domestic livestock. As a solution, Bruce and his colleagues proposed that the ranchers time their cattle drives to occur during the dry season. At first this advice helped, and it looked as though the government might help regulate cattle drives. But the solution didn't hold for long, thanks to the ever-volatile affairs of human politics. The 2007 presidential election ignited explosive violence between the two largest ethnic groups in Kenya: the Luo and the Kikuyu. More than a thousand people died, and hundreds of thousands of people were displaced from their homes due to post-election violence. Protecting national parks and wildlife became a much lower priority for the government. The cattle barons soon took advantage of the lack of scrutiny and drove hundreds of thousands of cattle into the Tsavo region, where they remained throughout the year. Once again, lions were treated as pests to be exterminated. Today the conflict between the cattle barons and lions continues in Tsavo and all across Africa.

Biologists working in Africa are concerned not only about the lions, but about other mammal species as well. Many large mammal species in East Africa are particularly vulnerable. In recent decades elephants have been poached for their ivory tusks, and rhinos have been heavily poached for their horns. In 2007 I was able to see some of the effects of this on my trip to Kenya and the enormous Tsavo National Park system of over 9,000 square miles. I saw many majestic elephants in Tsavo, and they did not shy away from my jeep. But there were no live rhinos to be seen anywhere. When I visited the Tsavo East National Park headquarters, I was shown vast storehouses of rhino and elephant skulls. The skulls were collected by the Kenyan Wildlife Service as remnants of poaching and drought between 1968 and 1971. After the early 1970s, poaching problems rapidly increased. From 1972 to 1988, the population of elephants in Tsavo dropped from 25,000 to 5,000. An international ban on ivory was put in place in 1989, and the number has since recovered to just over 11,000. Challenges for rhinos are much greater and their recovery much less likely. In the late 1960s, there were between 6,000 and 9,000 black rhinos in Tsavo. By 1989 that number had dropped to fewer than 30. There are now a couple of small highly protected areas in Tsavo where black rhinos have been relocated in hopes of saving the species from extinction, but poachers are bold, desperate, and well-armed. They will risk almost anything for horns that today sell for over a thousand dollars an ounce (or $300,000 per horn). Park guards have the right to shoot poachers on sight; but the slaughter continues. The job of trying to save these species from extinction is extremely challenging.

The lions of Tsavo and the three Pattersons represent important parts of the Field Museum's history. Bruce's story, in particular, brings out an important application of curatorial research that has become of great importance today: conservation. John Henry Patterson was from a period where his goal was to hunt lions (together with a sizable quantity of other game animals) and not give much thought to their conservation or survival as a species. In contrast, Bruce Patterson is part of a generation that is involved with conserving lions—as well as elephants, rhinos, and other particular species sometimes referred to as flagship species—because even as single species they have the ability to raise support for conservation efforts in general. And the conservation efforts of natural history museums today go well beyond concerns for flagship species. They also are attempting to protect entire ecosystems.

(*Top*) John Henry Patterson with the first of the man-eaters that he brought down. (*Bottom*) The rickety tree platform that Patterson sat in night after night waiting to ambush the second man-eater and almost becoming lion food himself. Both photos taken in 1898.

While on museum business in Kenya in 2007, I visited the man-eaters' den in Tsavo National Park that had been discovered by John Patterson more than a century earlier. I took this picture of my colleague Chap Kusimba and our armed guard (for protection against possible lion attack) just before the three of us entered the cave. Inside it was cool, dark, and silent, giving us an appreciation for what Patterson must have felt when he discovered the cave littered with human bones.

Bryan Patterson, son of John Henry Patterson. (*Top left*) In 1946 while curator of paleontology at the Field Museum doing fieldwork in Texas. (*Bottom left*) In 1967 after becoming a professor at Harvard, comparing arm bones of a modern human with the 4.1-million-year-old elbow fragment of the "Kanapoi hominid" (*arrow*) that he discovered in Kenya.

(*Top*) Tsavo male lion sunning himself in September 1998. (*Bottom*) Tranquilized male lion being fitted with a radio tracking collar in 2007. The purple spots are Betadine antiseptic solution applied where tissue samples were taken for DNA analysis. *From right to left*: Field Museum curator of mammals Bruce Patterson (project leader), Alex Mwazo Gombe (Tsavo resident, project assistant, and student of Bruce's), and Simon Wanjohi (Tsavo resident and Earthwatch driver).

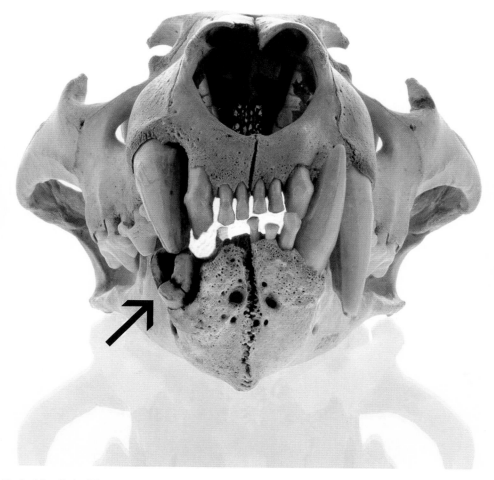

Skull of the first of the man-eaters killed by John Patterson in 1898, from the Field Museum's mammal collection. More than a century after its death, Bruce Patterson and colleagues determined that the reason for the lion's preference for human prey may have been the result of tooth problems. A broken lower right canine (arrow) infected with a root-tip abscess would have made it extremely painful to seize and hold large struggling prey like zebra and buffalo. Humans would have been softer, slower, and easier prey.

An enormous bull elephant that cautiously greeted my jeep in Tsavo National Park in 2007. This was one of hundreds of these majestic creatures I saw during my time there. Their numbers have dwindled sharply due to poachers hunting them for their large ivory tusks. Since the international ban on ivory took effect in 1989, their population has recovered to about 10,000 individuals, but poaching still takes a heavy toll in many regions.

The only sign of rhinos I saw in Tsavo in 2007 were warehouses full of skulls assembled by the Kenya Wildlife Service at the Tsavo Research Centre. They are the remains of rhinos killed by drought and poaching during the 1970s and 1980s, driving the Tsavo population of black rhino down to only a few individuals by 1989.

Today it is not just flagship species that need conservation; it is entire ecosystems. Commercial logging has led the way in destruction of rain forest since the 1800s. This 1993 picture shows logging on the island of Mindanao in the Philippines. Over the last several decades, the efforts of curators and other scientists have helped slow the destruction of these dwindling habitats. But the challenge to stop or reverse the trend remains.

13

Saving the Planet's Ecosystems

Conservation of our environment is one of humanities' most challenging issues in the twenty-first century. It has increasingly become a mission of natural history museums to advocate for and help facilitate the conservation of our biosphere. We do not know precisely which species and ecosystems are keystone elements of the pyramid that allows human society (and even human life) to survive. Our scientific research of the deep past tells us that there is a tipping point at which extinction of certain species and ecosystems leads to a chain reaction of other extinctions, sometimes eliminating the majority of all species on the planet. This proves especially lethal for those species higher up on the food chain. Because of climate change, habitat fragmentation, pollution, and invasive species, thousands of species are threatened globally. Organisms are currently disappearing faster than they can be cataloged and understood. If we as humans want to remain as a part of Earth's long-term future, we need to conserve enough of the current ecosystems to support our presence. We need to adopt an altruistic longview approach and realize that we exist on this planet only as part of a large interacting network of codependent organisms.

Major natural history museums have special credibility in addressing conservation issues. This is due in part to their having vast biological collections with data on when, where, and how each specimen was collected. Many of these collections have been strategically assembled and organized by curators for centuries, and today they form a unique and irreplaceable library of the diversity of life on Earth. Empirically, there is no better identification key to the species of the world than the preserved plants and animals that were originally used by scientists to study, describe, and accurately identify those species. Because each cataloged specimen has archived data documenting the time and place it was collected, we can also track changes in biodiversity through time for many areas of the planet.

Another asset museums have for conservation work is its staff of curatorial experts in the biological sciences, including some of the world's leading biodiversity specialists. These scientists build the collections through their fieldwork. They do research on these specimens to describe new species and determine their place in the evolutionary and ecological web of life. They publish their research in scientific journals and monographs, building a deep base of knowledge about biodiversity and ecology. They are also committed to training the next generation of biodiversity specialists.

One Field Museum scientist who has been doing research in the rain forests of the Philippines for more than three decades is curator of mammals, Larry Heaney.[1] During this time he has built a highly effective network of Filipino colleagues, government officials, and indigenous peoples to identify and preserve critical pieces of the Philippine ecosystem. His team has gone into remote forested areas where they have discovered dozens of previously unknown species of animals and plants endemic to the Philippines, single islands of the Philippines, or even to single mountaintops (endemics are species that occur naturally in only one place on Earth). Larry and his international collaborators have used these discoveries to convince the Philippine government to establish several national park reserves within the rain forest. In these areas the land is now protected from logging, mining, and other human activities detrimental to the environment.

I have known Larry far longer than any of my other curatorial colleagues. He was a fellow student of mine in that fateful class of Bob Sloan's at the University of Minnesota almost four decades ago, when my university trajectory changed from business school to evolutionary biology and geology (and when Larry and I both still had long, dark hair). At the time neither one of us knew we would one day end up as museum curators in Chicago. In the fall of 1988, he started as a curator at the Field Museum, just five years after I had started there. Since coming to Chicago, Larry has become a strong curatorial presence in the Field Museum's efforts toward environmental conservation.

The Philippine rain forest is home to one of the greatest concentrations of endemic species anywhere on Earth, dwarfing the Galápagos, Hawaii, and even Madagascar. Over 60 percent of its vertebrate species are found nowhere else on Earth, including 70 percent of its mammal species. About 50 percent of its 11,000 plant species and 70 percent of its 21,000 insect species are also endemic. In some areas of the Philippines, a single mountain can have five or more locally endemic mammal species. But because of widespread habitat destruction from logging and mining, the Philippines biota is one of the most severely endangered in the world. Today less than 8 percent of the original rain forest remains.

One of the dozen or so national parks in the Philippines that Larry and his colleagues have helped establish is a 100,000-acre park around Mount Tapulao in central Luzon. Mount Tapulao rises high into the clouds, to about 6,700 feet above sea level. This lush mountain rain forest is a treasure trove of species richness. It is also a critical watershed for the island of Luzon and receives over 200 inches of rainfall annually, much of which arrives torrentially during typhoons. The Mount Tapulao rain forest contains the headwaters of major rivers that ultimately help fill the growing demand for clean, reliable water to the lowlands near Manila, a city of 12 million people. Protecting these forests has a very practical impact. The mountain forests act as a giant sponge, soaking up water during storms and releasing it gradually during the year. This averts flood damage in the rainy season and provides water throughout the dry season. The loss of rain forest and its corresponding watershed elsewhere in the Philippines has led to large-scale human disaster. In 1991 after a watershed on the island of Leyte was logged and converted to sugarcane plantation, the city of Ormoc suffered a flood that killed over 7,000 people. Sometimes you must protect nature in order for it to protect you.

The greatest challenge faced by Larry and his colleagues in conservation of the Philippines rain forest is the ever-growing human population. Although the entire land mass of the Philippines is only about the size of Arizona, it is inhabited by over 100 million people. The island of Luzon alone has 42 million people including over forty different ethnic groups of indigenous people, each with its own language and customs. Many of the ethnic groups living in Luzon still remain culturally distinct from the majority of Philippine society, retaining most of their historical traditions. Larry's team works closely with the indigenous Aeta people as part of the holistic community engagement necessary for his work to be successful. The Aeta are a dark-skinned people of small stature and frame, seldom exceeding five feet in height, who are believed by some anthropologists to be among the earliest human inhabitants of the Philippines. There are only a few thousand of them left today, and they often still live in houses of bamboo and cogon grass. They receive little protection from the government, and from birth have an average life expectancy of only about seventeen years (due in part to high infant mortality). But they choose to live independent of modern society, in relative harmony with the rain forest. They are an important cultural component of Larry's network of local collaborators. Modern society, on the other hand, poses a serious threat to the environment of the Philippines from major corporations vested in logging, mining, and large-scale farming. Like so many other places around the world, the Philippines is an arena for the struggle between nature and human exploitation. Larry and his colleagues are tasked with balancing the long-term conservation strategies for the Philippine ecosystem with the short-term economic needs of its people.

Larry's career-long research program—collaborating with the research programs of geologists, anthropologists, and other biologists—has led to a deep understanding of the origin and uniqueness of the Philippine ecosystem. The high diversity of small endemic mammals in the Philippines is the result of the islands having formed far from other inhabited land masses. The islands did not become stable land areas above sea level until about 30 million years ago. By 20 million years ago, several islands of at least 400 square miles had emerged. As time went on, plants and invertebrates invaded the islands with seeds, insects, and other small organisms blowing in or floating in from afar. Eventually, a diverse tropical ecosystem of plants and invertebrates developed without any larger animals.

The first rodent-like mammals appear to have arrived on the islands about 15 million years ago (time estimate based on molecular studies of modern Philippine species). They must have floated there over the ocean on natural rafts formed by storm-uprooted trees or other vegetation from the Asian coast hundreds of miles away. These first immigrants faced a rich environment that was full of unoccupied ecological niches and free of predators. They evolved into a variety of different species with specialized anatomical modifications enabling them to fill available niches of the rain forest from the forest floor to the tops of the tallest trees. There were no shrews in the leaf litter of the rain forest, so one lineage evolved into dozens of different small shrew-like forms called shrew rats. There were no monkeys or tree squirrels in the rain forest canopy, so another lineage evolved into a variety of large tree squirrel-like forms called cloud rats. In 2006 Larry Heaney and colleagues published a paper concluding that more than sixty endemic mammal species were descended from just two ancestral species that had reached there between 8 and 15 million years ago. The Philippines is a cauldron of evolution—or perhaps better described as many cauldrons of evolution.

Most of the native mammals of the Philippines today live in the well-established forest areas. One of the most interesting and endangered groups that is endemic to Luzon Island of the Philippines is a specialty of Larry's: the cloud rats (pages 322 and 326). This group of rodents contains about nine described species on Luzon, as well as several species elsewhere in the Philippines, all thought to be descended from a common ancestor that arrived on Luzon about 15 million years ago. Cloud rats are nocturnal (night active) and live high up in the tree canopy. Most of them are very poorly known because of their scarcity and the relative inaccessibility of their habitats. They are slow-moving, docile animals. The larger species are hunted for food, which has driven them to near extinction. Protected parks are necessary for the survival of this group.

In addition to the wealth of endemic mammals, the Philippines has an amazing number of other unique species, including over 200 endemic species of birds (e.g., page 327). The endemic plants of the Philippines are also amazingly diverse, with over 6,000 species occurring nowhere else on Earth. Interactions between scientists and the Aeta people of Luzon are only beginning to document medicinal properties of plants there. Some have been found to be effective insect repellents, and others aid in the control of bleeding problems in humans. The beauty of Philippine plants is unsurpassed. There are over 140 genera and 1,100 species of orchids in the Philippines, and nearly a thousand of those species occur nowhere else on Earth.

When Larry began doing fieldwork in the Philippines in 1981, there were only about two dozen active Filipino researchers studying wild mammals, birds, reptiles, and amphibians of the Philippines. Speaking or writing about environmental issues in the Philippines during the regime of Ferdinand Marcos was likely to result in the author being harassed, intimidated, fired, or worse. After the People Power Revolution deposed Ferdinand Marcos in 1986, things began to change for the better. With a grant from the MacArthur Foundation in 1991, Larry established a training program for Filipino field biologists and helped create the Wildlife Conservation Society of the Philippines (WCSP). Over the years Larry has been active in training Philippine students to become effective conservation biologists themselves. He has also been working with his Filipino colleagues to establish the first Philippine National Museum of Natural History. At the time I wrote this, the new museum was scheduled to open in late 2016.

The coral reef of the Florida Keys National Marine Sanctuary is four miles wide and 150 miles long. It is the third largest reef ecosystem in the world. Rüdiger Bieler,[2] curator of invertebrates, has watched the reef's health decline sharply in recent decades. By some estimates, 90 percent of the live structural corals have died due to a variety of factors, including pollution and environmental accidents. In most of the reef areas, branching elkhorn and staghorn corals have crumbled back into the ocean floor, and the once-flourishing boulder coral heads have turned into lifeless structures that no longer sustain native marine life. The coral heads comprise a keystone component of the reef community, and some of the larger ones took 500 or more years to grow. Rüdiger recently joined forces with colleagues from Florida's Mote Marine Laboratory. Their project is focused on re-skinning dead coral reef heads with the living tissue of the same species that originally formed these massive structures over the centuries. Using a newly developed technique, their coral restoration team grows thousands of coral polyps in laboratory tanks. They plant the living polyps in a hair-plug-like fashion onto the surface of the lifeless coral boulders. Within a few years, the planted polyps multiply, fuse with their neighbors, and resurface the once-dead reef structures before they crumble away. The rejuvenated coral structures return to their former role as the "rain forest of the sea." Coral reefs support approximately 40 percent of all marine life on the planet, including a complex of interacting organisms ranging from microscopic life-forms to fishes, turtles, and marine mammals. Rüdiger's work is another prime example of how curatorial research and action can help sustain the health of the Earth's ecosystem.

Rüdiger has been a curatorial colleague of mine for almost twenty-five years, as has his wife, Petra Sierwald (curator of arachnids and myriapods, discussed in chapter 7). He has a particular interest in marine mollusks such as snails, clams, oysters, and their close relatives. He studies how they have evolved over time and how their diversity is impacted by human activities. Rüdiger has a dictionary of mollusks in his head and can identify thousands of species on sight during the hundreds of dives he makes surveying reef biodiversity. While his survey is not rapid, it is comprehensive. He has worked nearly three decades inventorying the Florida Keys and has recorded over 1,700 species of mollusks alone, tripling the number of species once thought to live in the region. Rüdiger's extensive inventory establishes a baseline species list for the ecosystem of the Florida Keys. It is used to assess the ecological damage of events ranging from ship groundings on the reef to major oil spills, and how best to maintain the reef for future generations.

The American Museum of Natural History and the Field Museum have also created major applied, non-curatorial programs in environmental conservation. Both programs are tied to using natural history museum scientists and collections. These action-based teams, including PhD-level scientists, were not conceived as formal scientific divisions with curatorial staff and basic research. Instead, they were designed to focus on immediate pragmatic objectives ("action") with clearly measurable results (e.g., acres of protected rain forest with enhanced level of indigenous community involvement in protection of their natural resources). The American Museum's priority was to focus on capacity building (helping countries or regions to develop the ability to achieve their own conservation objectives) while the Field Museum's priority was to protect measurable acreage on the ground. Although both programs were set up as non-curatorial divisions, each came to be through the efforts of a curator.

The American Museum's Center for Biodiversity and Conservation (CBC) in New York was one of the earliest efforts by a major natural history museum to create an applied conservation program.

Its goal was to bring the museum's extensive scientific and educational resources to bear on environmental conservation decisions and actions. It came to fruition largely through the organizational efforts of curator of paleontology and museum vice president Mike Novacek[3] with support from museum president Ellen Futter.

In the early 1990s, Novacek had been approached by Michael Klemens, a research associate in the department of herpetology and ichthyology. Klemens was an ecologist and a herpetologist who was promoting the idea of a major conservation program at the museum. Novacek had been looking for a way to connect science more effectively with environmental concerns and quickly took to the idea of building a conservation center within the museum. He assembled a task force headed by another curator, ornithologist Joel Cracraft, to develop a plan for the new program. Over a period of several months, Novacek and the task force helped transform the ideal into a significant program of the museum that today has over twenty staff members, most of whom are supported through grants from federal agencies, private foundations, and individual donors.

Novacek's interest in environmental conservation developed in the early 1990s as the result of his participation in a search for an ornithology curator. The short-listed candidates for the job were all required to give their public research presentations as part of the interview process. Several of the candidates reported that the bird populations that had first drawn their passionate interests to ornithology had sharply declined during the years they had been working on their dissertation research and were now highly threatened, endangered, or just gone. In his 2007 book *Terra: Our 100-Million-Year-Old Ecosystem—and the Threats That Now Put It at Risk*, Novacek relays his reaction to this dilemma:

> I was shocked by the wanton destruction, and by how it collided with the human need to discover what exists in the world around us. . . . [It is] disturbing to contemplate that in a few years we might fail to learn even the fundamental quality and quantity of that loss. We would then have an unprecedented mark on

history: we would be the generation that let life slip through our fingers.

For most of its history, the CBC directorship was been held by Eleanor Sterling,[4] who became director of the center in 1999. She held the position for fourteen years and developed the CBC into the dynamic program that it is today. Early target areas of the CBC have included the Bahamas, Bolivia, Madagascar, and Vietnam. Closer to home, the CBC launched an ambitious effort to assess what is known about the biodiversity of New York State, with rapid, or blitz, surveys in the New York area. Curators of the American Museum collaborate with members of the CBC on a wide range of environmental biology activities, including organizing international symposia, publishing technical papers, and training students, teachers, and U.S. Peace Corps volunteers.

Around the same time that the CBC became established in New York, Chicago's Field Museum developed an applied conservation program of its own. It began in late 1994 as the Environmental and Conservation Programs (ECP) alongside another organization focused on cultural issues, the Center for Cultural Understanding and Change (CCUC). In 2005 ECP and CCUC merged to become the division of Environment, Culture, and Conservation (ECCo), and most recently it became the Keller Science Action Center. The Science Action Center (as well as all of its previous incarnations) is focused primarily on the Andes-Amazon region of South America and Illinois.

Like the American Museum program, the Field Museum program was originally established by a curator of paleontology. Peter Crane (introduced in chapter 2) was curator of paleobotany and also vice president for academic affairs at the Field Museum in 1994. He saw the need and responsibility for natural history museums to connect their scientific assets more effectively with environmental concerns. As a botanist with expertise in living plants, he also had seen tragic habitat loss in various parts of the world. Furthermore, as a paleontologist he was familiar with global mass extinction events through time and the suspicion that due to human activities we may be entering one of Earth's greatest mass-extinction events. Peter worked closely with museum president Sandy Boyd to establish the ECP and convince the board of trustees that this would be a good move for the future of the Field Museum. He put together a budget and an initial organizational plan, and appointed Debra Moskovits[5] to head the office. With a strong background in tropical biodiversity, she had also been developer of the Field Museum's environmental and animal exhibits for the previous six years.

The new program of ECP began with Debra and two other half-time positions in 1995. Under her effective leadership, it grew into a powerful division of the museum that became ECCo and by 2011 included forty-six staff members.[6] ECCo's vision of conservation continues to translate the museum's science and collections into immediate action for conservation, cultural understanding, and the well-being of the people living in and around biologically rich areas, particularly in South America. ECP-ECCo also helped launch Chicago Wilderness (CW), an alliance of now more than 250 organizations working together to conserve biodiversity in the 6-million-acre Chicago metropolitan region.

The most prominent mission of the Science Action Center is to survey and understand biodiversity in ways that directly lead to protection of the environment. Two of the most effective ways in which the center does this are (1) to do rapid biological and social inventories of flora, fauna and culture, and (2) to produce easy-to-use freely distributed field guides for recognition of local plants and animals.

Rapid biological and social inventories (RSBI) are rapid surveys of the plants and animals that can be done quickly in poorly known and relatively intact habitats. Biodiversity experts are central to the RBSI field teams. Some come from the Science Action Center, while others come as consulting participants from the curatorial staff or from other institutions around the world. It takes anywhere from six months to a year to organize an RSBI. The first stage of implementation is for an advance team to helicopter, canoe, or hike into a remote area of rain forest to establish base camp. Soon afterward, larger scientific teams arrive there to conduct the survey of plant and animal species over the course of about three weeks. Occasionally the unexpected happens. Once, the rapid inventory group was raided by the Ecuadorian army, who had mistaken them for an encampment of FARC terrorists. They were overcome by twenty-two armed soldiers jumping out of army helicopters and wielding machine guns. Fortunately no one was seriously hurt. On another occasion, one of the helicopters that had been used to ferry the team into the rain forest crashed, killing its copilot. Luckily, none of RSBI team members were on board at the time. The group understands that it is all part of the risk for going into unexplored, unregulated territories. Such incidents have never dissuaded the RSBI teams from their mission, nor has the variety of tropical diseases, parasites, and difficult physical conditions that the team routinely encounters. Typically, the team has an impressive tropical disease résumé, with malaria, dengue fever, and leishmaniasis (mountain leprosy) being quite common. It takes a truly rare disease to impress this group. The teams do not attempt to produce an exhaustive list of species, nor do they attempt to do extensive research on specimens or name new species. Specimens thought to possibly represent new species are subsequently brought to the attention of curators and other biodiversity researchers around the world for later study. The goal of the surveys is to rapidly identify biological communities of outstanding significance that warrant immediate preservation.

Once the rapid part of an inventory has been completed, the teams spend a maximum of one year preparing a comprehensive report. The report is then presented to local and international decision makers, who set priorities and guide conservation action in the host country. By the end of 2014, the ECCo–Science Action Center group had conducted twenty-seven rapid inventories. They had helped establish seventeen protected areas in the Andes-Amazon region alone (focusing on Peru), totaling more than 25 million acres, plus another 7 million acres in other regions of the world (see map on page 337).

The Science Action Center has also produced hundreds of easy-to-use field guides online, which have received hundreds of thousands of visits from people around the world. These guides can be easily accessed on the Field Museum website.[7] Key to production of these guides is the Science Action Center's Scientist Robin Foster,[8] who is also an adjunct curator of botany and was part of the action team during its founding in 1994 as ECP.

Robin was hired for the ECP because of his unsurpassed expertise in plant identification and tropical ecology. He would be working mostly in Peru, a country in which a patch of forest the size of a football field had more native tree species than all of North America combined. Robin initiated a program to create tools for identification of plants in the field using his great knowledge of tropical flora and imagery. There were very few guides for the tropical Andes region available before Robin started making them. Over his years at the Field Museum, Robin led the botanical part of twenty-four rapid inventory expeditions including eighteen to the tropical Andes and Amazon region. He also put over 500 different rapid field guides on the Field Museum website, most of which were produced by him and his team. The individual guides range from a single page with 20 photos to 74 pages with 1,440 photos. They are simple, easy-to-use diagnostic tools assembled for a broad audience, ranging from scientists to indigenous people of the rain forest. They can be downloaded from the Field Museum's website at no charge, and they are currently being downloaded at a rate of more than 150,000 per year. The guides focus on clear color imagery and names rather than a lot of text. Robin reasoned that today detailed text can be googled once you have identified the species. Robin was the most productive biodiversity expert in the ECP and ECCo divisions, and one of the most productive scientists in the museum. He retired in 2013, but he remains an adjunct curator of botany and a participant in the Science Action Center.

The Science Action Center also includes social scientists, and one of the most important is curator of North American anthropology Alaka Wali.[9] Alaka was hired in 1995 to be director of the Field Museum's Center for Cultural Understanding and Change and curator in the anthropology department. Her part in the Science Action Center's conservation program has been to identify local ecological knowledge and cultural practices useful for developing conservation strategies. Native people have special knowledge of their environment and often already have measures in place for protection of their resources. Alaka and her group of culturally sensitive social scientists identify local peoples and activities that can be integrated into the protective master plan for the region. This gives conservation efforts long-term stability after the RSBI team leaves. Working closely with biologists on the team also adds depth to Alaka's anthropological research, because it enables her to better understand the plants and animals that are critical to local cultures.

The Field Museum's Keller Science Action Center has been successful in rapidly producing conservation results, particularly in the establishment of protected parks in the Andes-Amazon regions of South America. In parallel, the curatorial research programs in the Integrative Research Center have provided a more in-depth understanding of biodiversity and extended the global footprint of the museum's conservation efforts. Integrative collaborations between the research center and action center have the potential of great synergy. An ongoing challenge of institutional administrators is to find the right balance and interactive mix between the two centers for maximum effectiveness.

Environmental conservation today has come down to a matter of ecological triage. We have to make hard choices and ask hard questions, because our resources to protect biodiversity are limited. Where and when do we prioritize single species (e.g., the lion and the black rhino)? Where and when do we instead prioritize particular areas (e.g., the Andes-Amazon region and the Philippines)? Where and when do we prioritize particular environments (e.g., the rain forest and coral reefs)? Which species and ecosystems are most important to humanity's survival? Curators have a critical role here, because biodiversity research will always be necessary to form a knowledgeable base from which to strategize conservation actions.

In the words of Baba Dioum, founding member of the International Union for Conservation of Nature:

> In the end we will conserve only what we love. We will love only what we understand.

Curator of mammals Larry Heaney in the high mountain region of Luzon, holding a giant bushy-tailed cloud rat (*Crateromys schadenbergi*). This endangered species occurs only in the Philippines, high in the tree canopy of the mossy cloud forests.

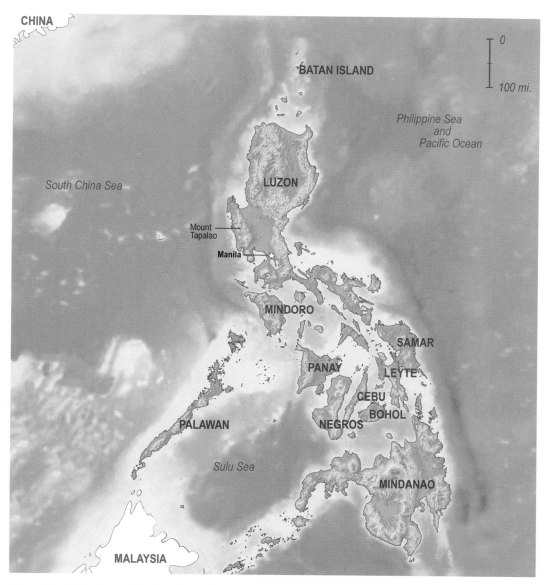

Map of the Philippines with major islands identified in blue. Most of the more than 7,000 islands that make up this Southeast Asian country are too small to be seen on this map, and their collective land mass is only about the size of Arizona.

Mount Tapulao National Park in central Luzon of the Philippines with over 100,000 acres of protected rain forest, first established in 2004. This is just one of over a dozen protected rain forest sanctuaries that Larry Heaney and his colleagues have helped establish over the last twenty years.

(*Top*) Larry working in the field with Filipino colleagues Danilo Balete, Ricardo Buenviaje, and Joel Sarmiento to identify a group of small Luzon mammals. (*Bottom*) An Aeta hunter from the rain forest of Luzon who has acted as a guide for Larry's research team. The Aeta are indigenous people thought to be among the earliest inhabitants of the Philippines.

Cloud rats endemic to Luzon Island of the Philippines. (*Top left*) The tiny Luzon tree mouse, *Musseromys gulantang*, weighing about a half ounce; (*top right*) the bushy-tailed cloud rat, *Crateromys schadenbergi*, weighing about three pounds; (*bottom left*) the slender-tailed giant cloud rat, *Phloeomys pallidus*, weighing six pounds, measuring about 30 inches long; (*bottom right*) the short-footed tree rat, *Carpomys melanurus*. These and five other species of cloud rats are thought to have descended from a common ancestor that reached Luzon at least 15 million years ago.

A few of the over 200 endemic bird species unique to the Philippines. (*Top left*) The Philippine frogmouth, *Batrachos septimus*, a bird that builds its nest from its own downy feathers held together with moss, lichens, and spider silk; (*top right*) the trogon, *Harpactes ardens*, one of the world's most colorful birds; (*bottom left*) the ecologically threatened rufous hornbill, *Buceros hydrocorax*; (*bottom right*) the Philippine eagle, *Pithecophaga jefferyi*, a critically endangered species that is also the national bird of the Philippines.

A few of the more than 6,000 endemic plant species unique to the Philippines. (*Top left*) One of over 900 species of endemic orchids; (*top right*) globular inflorescence of the jade vine, *Macrobotrys*, which is pollinated by bats; (*bottom left*) *Nepenthes merrilliana*, a carnivorous plant that feeds mainly on insects, but also occasionally on mice, lizards, and small birds; (*bottom right*) the giant corpse flower, *Rafflesia verrucosa*, named for its scent of rotted meat which attracts the flies that pollinate it. The flowers of this genus are the world's largest and can exceed three feet in diameter and weigh over 20 pounds.

Curator of invertebrates Rüdiger Bieler, looking at hundreds of very young coral polyps that he and his colleagues have raised in captivity at Florida's Mote Marine Laboratory. He and his team plant these living polyps on acres of dead coral boulders in the Florida Keys National Marine Sanctuary to help revitalize the barrier reef.

Rüdiger Bieler in the Florida Keys National Marine Sanctuary, chiseling holes into the boulder-like skeletons of dead brain corals. He then inserts live coral polyps into these holes to help eventually bring this part of the reef back to life (see next page).

(*Top*) Rüdiger Bieler and his colleague insert live coral polyps into the chiseled holes of the dead coral head. In about five years, the coral boulder should once again be covered with a skin of live coral polyps, helping to restore health to that region of the reef. (*Bottom*) A healthy section of the Florida barrier reef.

Director of the American Museum's Center for Biodiversity and Conservation (CBC) Eleanor Sterling with colleagues on an expedition to the Na Hang Nature Reserve of northern Vietnam in 2013. *From left to right*: Mary Blair (CBC), Eleanor Sterling, and students Duong Thuy Ha and Nguyen Van Thanh.

Vice president for the Field Museum's Environment, Culture, and Conservation Division (ECCo), Debra Moskovits (in the light blue shirt) in a dugout canoe in the upper Amazon region of Peru in 2009. Behind her is Corine Vriesendorp (director of ECCo rapid inventories and conservation tools), and at the ends of the canoe are two Maijuna indigenous people from the community of Nueva Vida.

Rapid inventory team helicoptering into the remote Kampankis Mountain region of the Peruvian rain forest in 2011.

Conservation ecologist and adjunct curator of botany Robin Foster (*second from right*) with students in the rain forest of Peru using his field guides on a rapid inventory exercise.

Applied cultural research director and curator of anthropology Alaka Wali (in blue and white shirt) with three members of the Shipibo people, in the buffer zone of Cordillera Azul National Park, northern Peru.

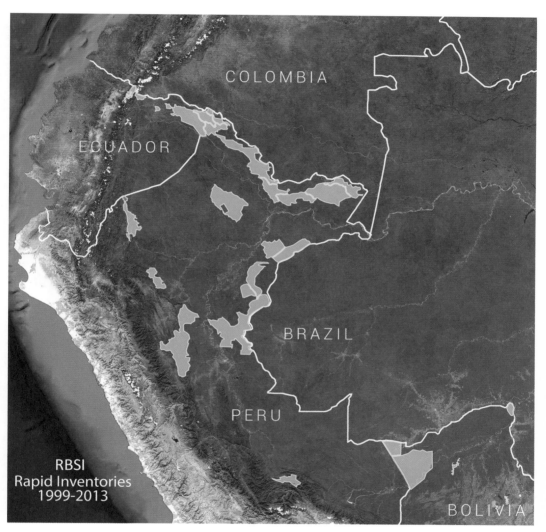

Areas of the Amazon headwaters that have been inventoried by the Field Museum's Rapid Biological and Social Inventories program (RBSI) since 1999 (*areas shaded in light green*). These inventories have served as a foundation for regional and national governments to establish seventeen protected areas encompassing 25 million acres. Today natural history museums are making a real difference in environmental conservation.

Twilight and a few minutes after, as seen from the front of my tent in southwestern Wyoming.

14

Where Do We Go from Here?

It was twilight on June 30, 2014, in southwestern Wyoming, and I was starting to relax on a high mountain butte just above the fossil quarry. My field crew, students, and I had arrived from Chicago earlier that day, and we had just finished setting up camp. As the fiery-orange sunset faded to a clear, dark sky filled with stars, I remember feeling the silence of the cool mountain desert. The students were eager to begin their first season of fieldwork digging for fossils the next morning, while I was contemplating starting my fortieth season of doing the same. The quiet of the evening put me in a reflective mood. I had come a long way since my first undergraduate class in paleontology at the University of Minnesota, yet I could still remember it clearly. The layers of experiences I have accumulated over time as a student, a curator, and an administrative leader have helped me form a perspective on my profession that I could not have had at any earlier point in my life. On that dusty butte in Wyoming, I reflected on the importance of natural and cultural history museums in human society. I reflected on the course of my career and what had helped make it a successful one. I thought about the unique importance of basic research and its applications.[1] I also thought about how the job of curators and museums might evolve to address growing challenges.

Natural and cultural history museums serve unique roles in society. They safeguard vast collections of cultural and biological heritage that could never be duplicated. The collections, together with full-time research curators, give museums the ability to answer some of the biggest questions in science, and empirically document those conclusions with specimens and artifacts. The diversity of collection strengths and researchers gives museums the flexibility to opportunistically address the ever-changing interests of society, even as issues become more global in scale. But as museums entered the twenty-first century, many began to experience growing pains with regard to maintaining enormous collections and diversely talented research staff.

Collections have expanded until storage infrastructure is bursting at the seams and in need of costly expansion. The last time that the Field Museum expanded its collection storage space between 2001 and 2004, the construction costs came to $86 million, not including the resulting annual increase in maintenance costs. New methods of research result in the need for new and expensive types of specialized collection storage. In 2005, for example, the Field Museum started a cryogenics facility for storing frozen tissue samples at −310°F for DNA analysis. Today this collection contains over 300,000 samples that must be kept in liquid-nitrogen-cooled tanks in perpetuity.

The cost of supporting optimal research staff has been rising as well. As collection size and breadth expands, curatorial coverage of the diverse scientific fields becomes unequal, with gaps in some areas and redundancies in others. When I wrote this book, for example, there were two curators of birds, two curators of mammals, but no curators of fishes, reptiles, or amphibians at the Field Museum. Also, because many major museums have existed for more than a century, the social issues addressed by them and their research staff have evolved. Today, for example, conservation issues are more of a mission priority than they were fifty years ago. As each museum grows and addresses broader issues, the costs of doing so rises. What is the key to meeting these ever-growing challenges?

In the future, natural history museums will have to enhance inter-institutional collaboration and integration. Partner universities will also play a role here, particularly with things like the rising costs of scientific journal subscriptions and housing resident graduate students in museums. A specialized consortium of U.S. natural history museums and partner universities (possibly even including major zoos, aquaria, and botanic gardens) could be formed for strategic planning of collection, research, and educational objectives. This has already begun in other countries (e.g., the Alliance of Natural History Museums of Canada). The new U.S. consortium should include a core of the big three natural history museums (the American Museum of Natural History, the Smithsonian National Museum of Natural History, and the Field Museum) plus dozens of other museums around the country. Keeping this as a national effort could facilitate coordination with federal funding agencies and foundations. A natural history museum consortium could more efficiently use resources by streamlining inter-membership loans and exchanges, coordinating collection growth and curatorial hires among member institutions, and producing traveling exhibits to rotate among member institutions. The consortium could standardize collection digitization efforts of all member institutions to form a national database of collection images and information available online. Even the applied conservation programs of member museums could benefit from closer, innovative collaborations with one another. The consortium might also coordinate ways to handle one of the biggest challenges in the United States today: the growing problem of science illiteracy.

Through the twentieth and into the twenty-first century, the robust growth of academia had many outcomes, most of which were positive. The U.S. academic system became one of the world's best in the area of research and higher education (university through postdoctoral). In fact, today the United States is still the number one university destination for international students from around the world. There were 819,644 foreign students in U.S. university programs for the 2012–13 academic years, according to *U.S News & World Report*. But ironically, there has been a growing domestic problem with science illiteracy that has developed in parallel with the expansion of higher-education excellence. The American *primary* educational system and educational efforts targeting the general public have not kept up. As a result, there has been an increasing divide between the American academic culture and the bulk of society. How did this happen? In part, the focus on academic prominence at the highest level was a victim of its own success, especially for the sciences.

In order to meet the competitive requirements of scientific publishing, grant success, and promotion, researchers of my generation had to focus on producing publications that targeted the professional scientific community.[2] This is because that professional audience is the same one that makes up the peer review community for papers submitted to scientific journals, and it has the power to accept or reject them. Furthermore, the same peer community makes up the external review panels for grant proposals submitted to the National Science Foundation, the National Institutes of Health, and other federal granting agencies. The peer community also acts as outside advisors for decisions on granting tenure and promotion. Thus, a successful academic research career in the top museums and research universities is dependent on effectively networking with other professional scientists to expand and promote one's own research abilities. The end result of all this was that my contemporaries and I were taught to write to an audience consisting of other scientists and graduate students in our publications. The interconnected system of research, publications, grants, and the scientific community created a culture whose influence and productivity was much greater than the sum of its parts.

Scientific discovery, productivity, and efficiency rose as a result of this selective interconnectivity. Most of the country's greatest scientists were producing a record number of publications for other scientists and university students. Dissemination of scientific advances among academics reached an all-time high. But efforts to communicate the nature and importance of modern science to the general public did not keep up accordingly. The more-or-less "trickle-down" method of science education in the United States during the late twentieth and early twenty-first centuries stopped short of reaching to the primary school levels and general public in an effective way. Scientific illiteracy grew rapidly.

In 2000 the Program for International Student Assessment (PISA) found that the United States ranked 15th among 41 industrialized countries of the world in science proficiency among fifteen- and sixteen-year-olds. By 2009 this ranking dropped to 23rd out of 74 countries, and by 2012 our ranking had dropped to 28th out of 65 countries. In 2008 a report by MIT found that 216 million Americans were scientifically illiterate. In a country that prides itself in being number one in so many respects, these are alarming figures. A 2014 poll by the National Science Foundation found that one out of four adults in the United States did not know that the Earth orbited the sun, and fewer than half knew that humans evolved from earlier species of animals. Forty-five percent of the adults sampled believed that astrology is a science. Even in China, where astrology originated sometime around the third century BC, only 8 percent believe it is a science. Today much of the general public in the United States depends on politicians, talk radio, relatives, or religious counselors rather than scientists to help them decide important issues about the natural world. Such issues include the causes of and appropriate response to global climate change, the use of molecular research in fighting disease and world hunger, and whether "creation science" should be taught in the biology classroom.

As a professional biologist and paleontologist, one of my particular concerns has been the anti-evolution sentiment building in the United States over the last several decades. This is of particular relevance to major natural history museums because so many curatorial research programs involve the evolution of biodiversity. According to a 2014 Gallup poll, 42 percent of the U.S population still believes that the universe is less than 10,000 years old, that all species were created essentially at the same time, and that we should consider the book of Genesis to be the default truth held by all. There are now huge Creation museums around the country who portray themselves as "science" institutions, complete with dinosaur exhibitions to entice the younger crowd. I have been through the Creation Museum in Kentucky, and it is remarkable. It has a large wing on Adam and Eve with a series of attractive dioramas, all with full-size figures of Adam, Eve, the tree with the famous apple, and the snake of temptation. There is a realistic-looking diorama of young children playing with baby dinosaurs, and another huge exhibit hall implying that Darwinism led to the decay of Western civilization. There is a planetarium in the museum that explains why we can see stars that are billions of light-years away (the light supposedly goes through some sort of time warp). There was even a wing on how Noah built an ark out of logs that could hold two of every species on the planet, including all of the dinosaurs.

The problem with creationism isn't within a moral or religious context. Most of the people on this planet find comfort in one faith-based belief or another. The problem occurs when creationists misrepresent themselves as scientists (e.g., "creation scientists") in a thinly veiled effort to get specific religious beliefs taught in science classes. Henry M. Morris is considered by many to be the father of "modern creation science," and he is the founder of the Institute for Creation Research. In a 1974 book advocating creation science, Morris demonizes evolutionary biologists, writing: "Satan himself is the originator of the concept of evolution." Such statements are not science; in fact, they are anti-science. A few scientists, in turn, have responded by advocating pure atheism as a requisite of science, an extreme position that has failed to convince people of faith.

The controversy is partly one of misunderstanding and misrepresentation that could be minimized by looking at it in a different context. Religion and science are simply not comparable ideologies.

Creationism is a top-down religious *belief* that provides supernatural explanations for the patterns of biodiversity we see today. Science, on the other hand, is a bottom-up *method* charged with providing natural explanations for those patterns. We never really arrive at the "absolute truth" in science, because science continually challenges its hypotheses as information accumulates through exploration, experimentation, and new discoveries. Science is a type of journey, ever learning but never arriving at the ideal perfection of truth. The most critical issue here isn't even whether or not God is responsible for the existence of Earth and its inhabitants.[3] It is an issue of what constitutes science, and what constitutes theology. Science is responsible for proposing natural theories (e.g., gravity, evolution) to explain natural patterns (e.g., falling objects, hierarchical groups of organisms with uniquely shared characteristics). It is the job of science to work with observable phenomena and base its conclusions on hard evidence. Religious faith, by definition, does not require evidence. This is not a commentary on the value of religion or dictating truth in science (there are, in fact, many scientists who have some degree of religious faith or spiritualism). It is simply to clarify that the place for teaching religion, faith-based principles, or the meaning of life, is in a theology class, a church, or a philosophy class and not in the science classroom. It makes no more sense to teach creationism in a biology class than to teach biochemistry in a sermon on the church pulpit.

Part of the challenge for scientists in the future will be to help non-scientists better understand the job, function, and benefits of science. If the problem of science illiteracy is allowed to continue along its current trajectory, the result may be further erosion of national stature and even a decline in long-term economic health. In the past we have averted this to some degree by being a brain drain to the rest of the world for their top scientific minds. I remember decades ago when postdoctoral students would come to the United States from China and elsewhere hoping to stay here eventually for a career and a better life. Although this is based only on personal observations, I do not see that as much anymore. With shifting global economic balances and tightening immigration policies, this ability to siphon off the world's best and brightest appears to be eroding. China as well as other countries now offer much better incentives to their students in the United States to return home for their careers. And new challenges are growing that make immigration more difficult. The pressure is building for us to develop more scientific talent within this country, and that must start first with addressing the domestic problem of scientific illiteracy. Knowledge is power, and a nation with a knowledgeable public is a powerful nation. Perhaps we should even consider it an issue of national security. Ultimately it is science that combats deadly disease outbreaks, protects our electronic information networks, and helps us feed an ever-growing population. Science prepares us for global changes to climate, biodiversity, and even culture. Science also has the capacity to be used for dangerous things in today's highly competitive, politically divided world. Ignorance is not an option.

There is an increasing need for senior scientists to address the problem of scientific illiteracy. No one understands new science better than the researchers on the front line. As stated by the United Kingdom's Martin Rees, president of the Royal Society in 2006, "Researchers need to engage more fully with the public." Research is the primary part of scientific progress and should be the top priority of most every professional scientist. But research should be balanced with efforts to communicate its significance to the general public. This could help erode the cultural divide that exists between academia and much of the country's population today. As part of education reform, academic institutions in the United States and elsewhere should assist researchers who are willing to engage the public, and to give them more positive reinforcement for successfully doing so. In a 2008 study, research physicist Pablo Jensen and colleagues determined that popularization of science generally has little or no positive effect on a scientist's career. In fact, their study found that many scientists believe it to be *detrimental* to their career. In the study of the Royal Society mentioned above, 20 percent of the surveyed researchers said that public engagement activities could actually be barriers to their career advancement. This aspect of academic culture should change. The administrative powers within those institutions should provide clear encouragement to professors and curators who can contribute in this regard. As recommended by the Royal Society study mentioned above, researchers "need to be rewarded . . . for undertaking public engagement activities," and perhaps even the culture of professional scientists need to value such contributions from their colleagues more highly.

Museum curators are ideally positioned to address the problem of science illiteracy. Unlike universities whose sole educational responsibilities are to university students, museums by definition are charged with educating the broader public, from young to old. This gives curators both license and a forum to reach a much wider audience. Curators work with material and issues that are inherently popular with the public. They can work with exhibitions departments to create captivating exhibits explaining the substance and importance of their research and of science in general. They can engage the public using a citizen-science approach like I have done with amateurs and commercial collectors in Wyoming (chapter 3), like Willy Bemis has done at the Alabama Deep Sea Fishing Rodeo (chapter 5), like the staff of the Bird Division does in downtown Chicago (chapter 7), and like Bruce Patterson has done with Earthwatch in Africa (chapter 12). Curators can also use the interdisciplinary aspects of their work to address broad issues to the general public through books, magazine articles, and social media.[4] Younger curators will always need to focus on establishing themselves in their peer scientific community and building credibility and intellectual context. There is a ramping up of scientific experience and perspective in the early stages of any scientific career that often requires an undistracted focus. But more established tenured curators should increase their efforts to engage the general public.[5] Many have rich experiences that could be of engaging interest to the public.

My life as a curator has been filled with challenge and adventure, and rewarded with discovery and a sense of accomplishment. I feel fortunate to have had such an immensely fulfilling career. I believe that the ability to understand the natural and cultural history of our world, and to monitor those things as they change, is key to our survival as a species. Part of the curatorial role is to help the public appreciate this. There is a certain curiosity about nature and culture with which everyone is born. Some individuals lose this as they grow into adulthood, but it is an interest that can be awakened. At this stage of my career, I will write more toward broader audiences to encourage this awakening.

I will also try to nurture the curiosity that young students have about nature while they are still filled with the wonder of it. Encouraging new generations of potential scientists and passing on what we have learned is a key role for curators. One of my favorite ways of doing this has been through my "Stones and Bones" field paleontology course, especially during the two weeks that we are digging fossils in Wyoming. There is nothing like real fieldwork to demonstrate the thrill of exploration and discovery. Teaching an energetic young group of highly motivated students is invigorating. Their enthusiasm is contagious, even to someone like me who has been doing this for over four decades. It is also personally nostalgic. It reminds me of long ago, when I was a student imagining what might hopefully lie ahead for me as a professional paleontologist. May the world never run out of passionately curious people with extraordinary aspirations.

It takes tens of thousands of cabinets and shelving units to contain the Field Museum collections. Here you can see one of thirty-four aisles of cabinets needed just to house the fossil plant collection.

The last major expansion of the Field Museum's collection storage facility occurred in 2001–4 with construction of a two-story underground expansion to add 186,000 square feet at the cost of over $80 million. (*Top*) Excavation on the south end of the museum. (*Bottom left*) Museum president John McCarter (*right*) with chairman of the board, Marshall Field V, inspecting the progress of the project in 2003. (*Bottom right*) Excavation on the east end of the museum.

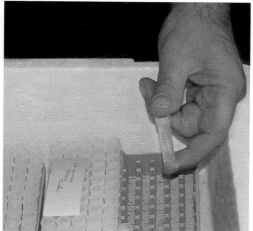

The cryogenics facility of the Field Museum (*top*) is where the frozen tissue collection has been stored since 2005 and now includes more than 300,000 frozen tissue samples representing tens of thousands of species. These collections are used for DNA analysis in a variety of research programs including the search for evolutionary relationships of plants and animals and the fight against disease. The small samples of liver, heart, or other tissues are stored in small vials (*bottom left*). Racks of the vials are stored in liquid-nitrogen-cooled tanks (*bottom right*) with a temperature of −310°F. Under these conditions, the samples could remain viable for genetic research for thousands of years to come.

Recruiting and training future generations of scientists will always be one of the most important responsibilities of natural history museum curators. Shown here is my "Stones and Bones" paleontology class of 2014, on Lewis Ranch in Wyoming. Front row includes the author (kneeling behind the orange-and-white power saw) and the rest of the museum field crew members. Behind the front row are the students. To my right in the red-and-white-checked shirt is Emily Graslie, host of *The Brain Scoop* on YouTube. She did a three-part story on us that summer (e.g., check online to see the Brain Scoop, "Fossil Fish, Pt. II: A History").

ACKNOWLEDGMENTS

I thank the colleagues whom I interviewed in an effort to get the facts straight. They were generous in their stories about their curatorial exploits. Any inaccuracies portrayed here are most likely the fault of the author. I also acknowledge all of the curators I have known whom I could not include here, but whose stories are deserving of mention. Maybe one day I will write a sequel. For reading all or part of early versions of the manuscript, I thank Mark Alvey, Steve Gieser, Anne Trubek, Erin DeWitt, Peter and Elinor Crane, Richard Lariviere, Lynne Parenti, and Neil Shubin.

For their especially strong support of the Field Museum's Collections and Research division during its most fiscally challenging years (2008–13), I thank The Negaunee Foundation, The Tawani Foundation, The Grainger Foundation, The National Science Foundation, The John D. and Catherine T. MacArthur Foundation, the University of Chicago, the University of Illinois at Chicago, Richard and Robin Colburn, J. N. Pritzker, John and Jeanne Rowe, David and Juli

Grainger, John and Rita Canning, Barbara and Roger Brown, Bill and Linda Gantz, Gail and Terry Boudreaux, Karen Nordquist, Bob and Charlene Shaw, Richard and Jill Chaifetz, and several anonymous supporters. One of the most rewarding experiences of my time heading the Collections and Research division was to see how much outside support there was for the C&R division's research operations, collections, and critical staffing.

Finally, I especially want to thank my wife, Dianne, for patiently reading and commenting on every page of this book, and not minding when a random thought about the manuscript would cause me to get up in the middle of the night to write it down before I forgot it. Her patience and love helped make it all possible.

NOTES, ADDED COMMENTARY, REFERENCES, AND FIGURE CREDITS

PREFACE

Notes

1. The number of specimens in any major museum is a rough estimate at best. The current official count of specimens at the Field Museum is 27 million, but it could easily be ten times that number and will probably have changed upward by the time this book is in print. Each museum counts its specimens in a different way, and the Field Museum uses a conservative approach (e.g., a lot containing over one hundred specimens of a fish or 500 specimens of an insect may get a single catalog number). Big numbers can be good promotional tools, so some institutions count every piece. Perhaps a figure more important than number of specimens is floor space devoted to collection storage. The Field Museum, which currently has over 1.3 million square feet of floor space in its building, devotes over half a million square feet to cabinets, shelves, and other storage devices for its collection. A single cabinet can contain as many as twenty-five drawers, and a single drawer or two-by-three-foot shelf may contain hundreds of specimens.

2. Linnaeus often led a life of controversy. Besides his trouble with the church, he sometimes bumped heads with government authorities. One of my favorite stories about Linnaeus took place in 1735 when

he was passing through Hamburg, where he met the mayor. The mayor showed Linnaeus a wonder that he thought was going to make him rich: a taxidermied specimen of a seven-headed hydra. The piece had been looted years ago from a church in Prague. The mayor had been trying to auction it off, and the bids had increased to a princely sum. He had already turned down a bid of 30,000 thalers from the king of Denmark. Linnaeus, being the consummate naturalist that he was, quickly recognized the piece to be a fake manufactured from jaws and clawed feet of weasels artistically stitched together with snake skins. Based on the history of the specimen, Linnaeus surmised that it had been manufactured by monks to represent the Beast of Revelation. After Linnaeus tactlessly made his observations public, the bids to buy the piece dropped to zero. Linnaeus was forced to leave town quickly to avoid the mayor's wrath. For much more on the life of Linnaeus, see the book by Blunt (2001) listed below.

References

Blunt, W. 2001. *The Compleat Naturalist: A Life of Linnaeus*. Princeton, NJ: Princeton University Press.

Bullard, G. 2014. "Government Doubles Official Estimate: There Are 35,000 Active Museums in the U.S." *IMLS Press Release*, May 19, 2014.

Imprey, O., and A. McGregor. 2001. *The Origins of Museums: Cabinet of Curiosities in Sixteenth- and Seventeenth-Century Europe*. Cornwall, UK: House of Stratus.

Kendall, P. 1994. "Museum VP Leaves the Field: Spock Changed Once-Stuffy Halls into a Showplace." *Chicago Tribune*, July 5, 1994.

Linnaeus, C. 1758. *Systema naturae per regna tria naturae: Secundum classes, ordines, genera, species, cum characteribus, differentiis, synonymis, locis*. 10th ed. Stockholm: Laurentius Salvius.

Moravcsik, J. 1973. "The Anatomy of Plato's Divisions." In *Exegesis and Argument: Studies in Greek Philosophy Presented to Gregory Vlastos*, ed. E. Lee, A. Mourelatos, and R. Rorty, 324–48. Assen, Netherlands: Van Gorcum.

Ogle, W. 1882. *Aristotle on the Parts of Animals: Translated with Introduction and Notes*. London: Kegan Paul Trench.

Rader, K., and V. Cain. 2014. *Life on Display: Revolutionizing U.S. Museums of Science and Natural History in the Twentieth Century*. Chicago: University of Chicago Press.

Singer, C. 1931. *A Short History of Biology*. Oxford: Oxford University Press.

Soulsby, B. H. 1933. *A Catalogue of the Works of Linnaeus in the British Museum*. 2nd ed. 2 vols. British Museum.

Woolley, L. 1929. *Ur of the Chaldees: A Record of Seven Years of Excavation*. Harmondswort, Middlesex, UK: Penguin.

Figure Credits

Page viii: Photo by Lance Grande. Taking a break from fieldwork in the high mountain desert of southwestern Wyoming, on Lewis Ranch, 2010.

Page xvii: Photographer unknown; Field Museum negative #GN78508.

CHAPTER ONE

Notes

1. Donn Rosen's PhD students from the late 1970s and early 1980s included Lynne Parenti and Richard Vari, who both became curators at the Smithsonian National Museum of Natural History in Washington, DC; Ed Wiley, who became a curator at Biodiversity Institute & Natural History Museum in Kansas; Guido Dingerkus, who became a curator at the Muséum national d'histoire naturelle in Paris; Darrell Siebert, who became a curator at the Museum of Natural History in London; and me, who became curator at the Field Museum of Natural History in Chicago.

Writing a book like this provides an opportunity to acknowledge people and friends who contributed to your ultimate success. When I first arrived in New York, the graduate student who best helped me learn the ropes of the city, the graduate program, and the complex of personalities that I would be interacting with was Lynne Parenti. Lynne was a local: born in Manhattan and raised there and in the New York City borough of Staten Island. She developed an early interest in natural history while exploring the salt marshes and decaying docks of the waterways separating Staten Island and New Jersey. We both had offices in the ichthyology department of the museum and Donn Rosen as our major advisor. She was sharply intelligent with a dry sense of humor, and she made me feel at home immediately. After receiving her PhD in 1980, she went through a lengthy series of postdoctoral and term teaching positions in Washington, DC, London, New York, San Francisco, and Chicago before becoming curator of fishes at the Smithsonian National Museum of Natural History in 1990. She led a transitory life on soft money (grants) for a decade, but in the end she got the job she had wanted since entering graduate school: curator of ichthyology in a major natural history museum. Today Lynne is one of the world's leading fish systematists and biogeographers. Over the years she has done extensive fieldwork in Pacific coastal regions of New Guinea, Borneo, Sulawesi, Singapore, peninsular Malaysia, Taiwan, China, Hawaii, Tasmania, and New Zealand, as well as Cuba. She is an award-winning author with over 100 scientific publications, and in 2003 she became the first female ichthyologist elected president of the American Society of Ichthyologists and Herpetologists.

2. In cladistics, uniquely shared characteristics (also called derived characters) include specific physical features, but not the absence of those features. For

example, the absence of leg bones in worms is not a derived character. After all, even a rock lacks leg bones. The *presence* of leg bones is the informative feature, and organisms that have leg bones form the cladistic group Tetrapoda (tetrapods). In cladistics, absences are not the same as secondary losses. For example, snakes are tetrapods because they have the beginnings of leg bones early in development. Their legs are lost in adults because they stop developing prior to the adult stages of growth and development. The body grows, but the limb buds remain as tiny vestigial remnants. The lack of legs in adult snakes is due to a secondary loss (i.e., not the same as the absence of legs in a worm or a rock), and therefore they are still tetrapods. The loss is a derived feature within the lineage of snakes, denoting a much smaller group within the Tetrapoda. There are also fossil species of primitive snakes that still have legs as adults, further indicating that loss of leg bones in snakes was part of the lineage's evolutionary history. This may seem an obvious point, but in the 1980s there was an opposing school of systematics called phenetics whose goal was to treat absence and presence characters the same and to use overall similarity as the basis for classification of organisms. It was another part of systematic controversy that once existed but today has all but disappeared.

3. Cladistic method can be used to look at the interrelationships of Earth's geographic regions through a research program called biogeography. Just as systematic biologists use patterns of shared anatomical characteristics to decipher patterns of relationship among species, historical biogeographers use patterns of shared endemic species to decipher patterns of relationship among geographic regions. The earliest ideas of continental drift were derived largely from distribution patterns of closely related plant and animal species. Many fossil species in Brazil appeared to be more closely related to species in eastern Africa than to species in western South America. Drifting continents eventually provided an explanation for those and other plant and animal ranges that seem disjunct based on present-day geography. Integrative studies of geology, geography, and biology clearly show that the evolution of Earth's surface has been intimately tied to the evolution of its plants and animals. Congruent patterns of Earth's geohistory and biohistory reflect a fundamental order of nature. In addition to biogeography, cladistic method has been used to study the evolution of language, culture, and other non-biological features (see, for example, Rexova, Bastin, and Trynta 2006).

4. There are two basic parts to evolutionary biology: pattern and process. Comparative analysis of anatomy (fossil or living) and molecular structure (e.g., DNA) provides data that is strictly *pattern*. When science discovers what appears to be a non-random pattern in nature, it attempts to provide a natural explanatory *process* explanation for it (i.e., theory). Much like the theory of gravity is the best natural explanation for patterns of falling objects, the theory of evolution is the best natural explanation for the patterns of species relationship based

on shared characteristics. In fact, evolutionary theory eloquently explains the countless congruent patterns in nature in a way that no other natural theory can. In order to intuitively enrich and expand the theory of evolution, we extrapolate observed processes in real time (e.g., parents producing children, each child producing children of their own, selective breeding experiments) into unobservable processes in geological time (species splitting into two species, and each of those splitting into two species). We can even put together evolutionary theories about transitions of functional anatomical structures (e.g., fins to hands, hands to wings, gill arches to jaws, etc.) based partly on observed embryological development (e.g., embryonic gill arch structures shared by fishes and humans that develop in different directions with growth). We apply supportive evidence from Mendelian genetics, embryological development, comparative anatomy, geographic species distributions, and the evolution of viruses in observable time to enrich the theory of evolution further. The result is a broadly encompassing process theory that can eloquently explain countless patterns that we see in nature. In the words of the famous geneticist and biologist Theodosius Dobzhansky, "Nothing in biology makes sense except in the light of evolution." Although the process side of evolutionary theory is a focus of many biologists, it is primarily the patterns of relationship among species that constitute testable hypotheses in evolutionary research. Evolution, in any substantial sense, takes so much time (more than the entire history of human observing) that we must study it based largely on its present-day results (cladistic patterns in nature). Deciphering the overall pattern of biodiversity is the fundamental task for evolutionary science.

5. The scientific use of the terms *hypothesis*, *theory*, and *law* are often confusing, even in scientific publications, so in this book I use what I believe is the most generally accepted meaning of these terms. A *hypothesis* is a proposed explanation for a relatively narrow set of phenomenon based on current evidence. It is an empirically testable (and falsifiable) explanation. An example would be a hierarchical pattern of relationship (cladogram) indicating species **A** being more closely related to species **B** than to species **C** based on a set of shared characteristics. The most general trees like the one on page 19 are relatively stable, but the trees of similar-looking, closely related species are subject to frequent change as new character information and species are discovered, or as existing information is tested with new analytical methods. A *theory* is a broad explanation with many interconnected parts that explain a wide range of phenomena. It is generally a more comprehensive explanation than a hypothesis and usually harder to falsify. It has been repeatedly confirmed through repeated observation. Theories are occasionally presented as though they were laws, although in the strictest sense they are not laws. Examples would be the theories of gravity, heliocentrism, relativity, and evolution by natural selection. Scientific *law* is a descriptive or mathematical account of how things always behave under specific conditions. The simplest example is if A = B and B = C, then A = C.

6. For example, starting with Hennig (1966) and, for its early history, Nelson and Platnick (1981).

7. For example, Simpson (1961).

8. The fossil record has major gaps in both time (stratigraphy) and space (geography). Major gaps lasting tens of millions to hundreds of millions of years are documented for entire orders of animals. For example, lampreys have thirty-eight living species today, but the youngest fossil species is 120 million years old with no fossil record between then and today. Hagfishes have twenty living species today but are represented in the fossil record by only a single species that is 300 million years old. The logical consequence of such gap durations is that many lineages of animals or plants in the fossil record may be much older or younger than indicated by their fossil record. You also cannot trust the fossil record for detailed geographic range information. Many extinct species are known by fossils from only a single rock outcrop, which was certainly only a fraction of their former geographic ranges. So we end up with a fossil record that we know is only a minute sample of where and when extinct plant and animal species existed through time. It has been estimated that there have been 5 to 50 billion species that are extinct today (e.g., Raup 1991). There are currently only about 250,000 known extinct species known as fossils (Prothero 1999). That extrapolates to less than 0.005 percent of extinct species known from the fossil record. Stratigraphic succession of species (i.e., a sequence of species moving up through the stratigraphic levels of rock) is of relatively little use in reconstructing ancestor-descendant relationships among species. In a given locality, stratigraphic succession of species may be the result of closely related species moving in and out of the area due to changing ecological conditions over time rather than evolutionary transformation. It may also be the result of misreading individual variation within a species (e.g., note the individual variation among domestic dogs, which all belong to a single species). Last, it may be cyclic change within a local population due to changing environment.

The reasons for the incompleteness of the fossil record are many, and the conditions for a dead organism to become fossilized are both rare and complex. First, an aquatic environment is usually needed (explaining the relative scarcity of well-preserved fossilized terrestrial species as compared to fishes and other aquatic species in the fossil record). When a plant or animal dies on land, its body usually decomposes completely or is torn apart by scavengers. Second, even in aquatic environments, rare combinations of water chemistry, oxygen conditions, and sedimentation rates are required in order for dead organisms to become fossils. If you dig deep into the mud of many modern lakes, you will find no fish bones. This is because the skeletons either dissolved chemically in the water or were consumed by scavengers and microorganisms. Third, at any particular location today, most time periods are no longer represented as sedimentary rock. Either the sediments that might have encased fossils never accumulated or the sedimentary rocks were later eroded away along with any fossils they might have contained. Finally, many extinct species that became fossils may never be discovered due to their inaccessibility.

Although the fossil record does not help us definitively recognize specific ancestral species, it still constitutes strong evidence supporting the theory of evolution at a general level. There is basic order to broad levels of evolutionary complexity. The oldest fossils are only single-celled organisms. Simple multiple-celled cell organisms are found in rocks much older than any rocks containing multiple-celled organisms with heads. Organisms with heads are found in rocks much older than any rocks containing animals with heads and limbs. Animals with heads and limbs are found in rocks much older than any rocks containing animals with heads, limbs, and hair. Animals with heads, limbs, and hair are found in rocks older than any rocks containing animals with heads, limbs, and hair and that walked on two legs. Many such examples of increasing general complexity exist in the fossil record. As Neil Shubin (2008) puts it, "If, digging in 600-million-year-old rocks, we found the earliest jellyfish lying next to the skeleton of a woodchuck, then we would have to rewrite our textbooks." We will never find a 600-million-year-old woodchuck: I stake my scientific reputation on that prediction.

9. In November 1981, at the height of Colin Patterson's advocacy for the cladistic revolution, he was invited by Donn Rosen to give a seminar on evolution and creationism at the American Museum of Natural History. The meeting was part of the regular Systematics Discussion Group series, which was the most provocative of the museum's seminar series. Just prior to giving his presentation to this combative group, which included a number of traditional paleontologists and biologists ready for a fight, Patterson had read an anti-cladistic paper in *Science* magazine by the famous evolutionary biologist Ernst Mayr. Patterson had become frustrated with the pace at which the field of paleontology was moving to incorporate phylogenetic concepts. Freshly invigorated, Patterson delivered several provocative statements in his presentation. He stated that the incessant search for specific ancestors in the fossil record by evolutionary biologists and paleontologists had set evolutionary biology back by decades. He provocatively asked his scientific audience, "Can you tell me anything about evolution, any one thing that is true?" He did not doubt the reality of evolution; he was merely challenging the audience of paleontologists and traditional evolutionary biologists to lay all their significant cards on the table. He argued that the traditional methods of specific ancestral species identification were largely non-scientific. He compared the identification of particular species as ancestors to pre-Darwinian creationism, saying it was not a research governing theory because its power to explain was only verbal. It was instead "an anti-theory that had the function of knowledge but conveyed none." Paleontologists, he said, were confusing *pattern* and scientific hypothesis with *process* and scientific theory. The debate that fol-

lowed was loud and vigorous. Although this was meant to promote vigorous discussion among the scientists in the room (which it certainly did), it also had an unexpected impact. There was an anti-evolution special creationist in the room with a hidden tape recorder. Special creationists believe that all species were independently created, following a literal interpretation of the book of Genesis. Many (the Young Earth creationists) believe that all of this happened less than 10,000 years ago over a period of six days.

Patterson's talk was transcribed and widely circulated among creationists, and it became one of the most widely quoted (contextually misquoted) scientific remarks in creationist literature. His name was used as evidence for numerous legal challenges to teach "creation science" in biology classes in parity with evolution. Patterson, they claimed, was one of the world's leading scientists criticizing the scientific status of evolution itself. As a devoted Darwinian and a scientist who firmly saw evolution as the only natural explanation for patterns of relationship in nature, Patterson was mortified by the episode. He spent months responding to masses of mail and requests for information about his views on evolution. In the 1999 edition of his book *Evolution*, Patterson lamented: "Because creationists lack scientific research to support such theories as a young Earth . . . a world-wide flood . . . or separate ancestry for humans and apes, their common tactic is to attack evolution by hunting out debate or dissent among evolutionary biologists. . . . I learned that one should think carefully about candor in argument (in publications, lectures, or correspondence) in case one was furnishing creationist campaigners with ammunition in the form of 'quotable quotes,' often taken out of context." He submitted official opinions relating to U.S. court cases debating the merits of teaching evolution in schools and continued to clarify his position publicly. The struggle increased his work to a punishing level, which is thought to have led to his first heart attack in 1984.

Things got better for Patterson with time. After he recovered from his heart attack, he continued to clarify his position with regard to creationism and to distinguish the business of science from that of more metaphysical and faith-based pursuits. As time went on, he finished a second edition of his book *Evolution*, and the dust settled for him within the scientific community. Paleontologists were increasingly coming to the conclusion that Patterson was right about the merits and limitations of fossils. He was no longer the outsider. In fact, he helped create a new mainstream. In 1993 Patterson was elected as a Fellow of the prestigious Royal Society of London (the equivalent of the National Academy of Sciences in the United States). In 1997 he was appointed as a foreign member to the board of trustees for the Smithsonian National Museum of Natural History. In October 1997 at the Society of Vertebrate Paleontology (SVP) meeting in Chicago, he finally received the Romer-Simpson medal that had reportedly been denied to him in the early 1980s. In his acceptance speech, he graciously apologized for the ruckus he had started with the creationists, but not for having been the one to point out that it was not the *age* of fossils that was relevant to resolving evolutionary relationships, but their *anatomies*. It was a gratifying experience for me to present the medal to him myself as a representative of the SVP because it was he who first set me on the path to becoming a professional paleontologist. Colin Patterson became known as "the greatest fish paleobiologist of the 20th century" (e.g., Bonde 2000).

Added Commentary on Gary Nelson and Donn Rosen

Gareth Nelson was born in 1937 to parents who were longtime midwesterners. He grew up in the Englewood region on the South Side of Chicago with two brothers and four sisters, and was brilliant from an early age. He never graduated from high school. At age fifteen he went right from his junior year at Chicago Christian High School to the University of Chicago. After two years at the university, he enlisted in the U.S. Army at age seventeen. He was sent to the top-secret Sandia Base in New Mexico to work on advanced weaponry and electronics. Sandia was responsible for fabrication, assembly, and storage of nuclear weapons for the U.S. Department of Defense. The secret base played a key role in the United States' nuclear deterrence capability during the Cold War.

After serving for three years, Gary returned to Chicago and worked as a technical writer for Allied Radio Corporation writing assembly manuals for electronics. After a year at Allied Radio, he spent a year in Paris, spending most of his time becoming fluent in French through a program connected to the Sorbonne and the University of Paris. Eventually, he decided to return to academia. He went to Roosevelt University in Chicago and graduated in 1962 with a Bachelor of Science degree. Over his years in Chicago, Gary had the opportunity to visit with the Field Museum's curator of fishes Loren Woods, who encouraged Gary to go to graduate school. As a result, Gary attended the University of Hawaii for a PhD. But it was not until his postdoctoral work at the Swedish Museum of Natural History and the British Museum of Natural History that he became intrigued by the phylogenetic method (later dubbed "cladistics" by one of its staunch opponents, Ernst Mayr). Gary first read about phylogenetics in a paper by a Swedish entomologist named Lars Brundin (1966). The Brundin paper summarized the phylogenetic method of German entomologist Willie Hennig and applied it to a monographic study of chironomid midge flies. This amazing publication on a lowly group of gnats helped change the world of systematic and evolutionary biology by influencing Gary, who in turn influenced Colin Patterson and Donn Rosen, and eventually changed the course of systematic biology and paleontology.

In the fall of 1967, Gary was hired by the American Museum of Natural History as assistant curator, joining Donn Rosen in the ichthyology department. At first his research was primarily on the descriptive anatomy of fishes, but gradually his focus shifted more toward systematic method, scientific philosophy, and, most of all, cladistics. This was a bold move for an untenured assistant curator because some of the older curators and associates of the museum who called themselves evolutionary systematists did not accept cladistics. In fact, they were violently opposed to it. These scientists were reluctant to give up their notions that evolution could be read directly from the fossil record, that ancestral lineages were readily identifiable just by looking at the fossil record, and that fossils were somehow more important than living species in evolutionary studies. There is much more to the debate than I can possibly include in this book, but eventually Gary was successful in leading a movement to bring cladistic methodology to the core of systematic biology. For a more detailed discussion of the controversy, see Hull (1988), and for a more technical discussion, see Nelson and Platnick (1981).

Donn E. Rosen was born in 1929 in New York City to Russian immigrants. He was a loyal New Yorker, even though he lived much of his later adult life with his family just across the Hudson River in suburban New Jersey (the bedroom community of many New Yorkers). He loved the American Museum of Natural History, and he was the first person to show me how vested curators can be in the education and professional training functions of their institutions. His association with the museum began at age eight when he joined the museum's School Nature League. At age fourteen, he started working as a volunteer/assistant in the Fish Genetics Laboratory of the museum headed by Myron Gordon. By the age of twenty-two, Donn had already presented an award-winning paper at an international scientific meeting, "Mechanism of Insemination in Poeciliid Fishes" (basically, how guppies have sex, and the anatomy of their male sex organs). His time in New York was interrupted in 1952 when he was drafted into the U.S. Army to serve in Korea. But after the war he returned to New York, now with the GI bill to help pay for schooling. By 1959 he had received the degrees of BS, MS, and PhD, all from New York University, and throughout his educational training he maintained a steady presence in the museum working with the fish collection. In 1960 when a curatorial position in the ichthyology department of the museum opened up, Donn applied and got it. He remained there as curator for the rest of his life.

While Donn was curator, he presided over a total reorganization of the fish collection that grew from 500,000 to 1.5 million specimens. He did fieldwork in Guatemala and Australia, and he produced a series of highly influential publications on the evolution, biogeography, and higher classifications of fishes. He was an engaging speaker, rarely using notes for presentations, and was elected president of the Society of Systematic Zoology in 1976. More than a successful field collector, researcher, and museum citizen, Donn enjoyed interacting with the graduate students he had recruited from around the country, and he was a paternal figure in the ichthyology department to both students and staff.

References

Bonde, N. 2000. "Colin Patterson: The Greatest Fish Paleobiologist of the 20th Century." Special issue, *Linnean Society* 2: 33–38.

Brundin, L. 1966. "Transantarctic Relationships and Their Significance, as Evidenced by Chironomid Midges with a Monograph of the Subfamilies Podonominae and Aphroteniinae and the Austral Heptagyiae." Fjarde serien, *Kungliga Svenska Vetenskapakademiens Handlingar* 11: 1–472.

Darwin, C. 1859. *On the Origin of Species by Means of Natural Selection; or, The*

Preservation of Favored Races in the Struggle for Life. London: John Murray Publisher.

Delson, E., N. Eldredge, and I. Tattersall. 1977. "Reconstruction of Hominid Phylogeny: A Testable Framework Based on Cladistic Analysis." *Journal of Human Evolution* 6: 263–78.

Dobzhansky, T. 1973. "Nothing in Biology Makes Sense Except in the Light of Evolution." *American Biology Teacher* 35: 125–29.

Eldredge, N. 2005. *Darwin: Discovering the Tree of Life*. New York: W.W. Norton.

Forey, P. L., B. G. Gardiner, and C. J. Humphries, eds. 2000. "Colin Patterson (1933–1998): A Celebration of His Life." Special issue, *The Linnean*, no. 2. London: Academic Press.

Gordon, M., and D. E. Rosen. 1951. "Genetics of Species Differences in the Morphology of the Male Genitalia of Xiphophorin Fishes." *Bulletin of the American Museum of Natural History* 95: 415–60.

Grande, L. 1980. *The Paleontology of the Green River Formation with a Review of the Fish Fauna*. Laramie: Geological Survey of Wyoming, Bulletin 63. [2nd ed., 1984.]

Hennig, W. 1966. *Phylogenetic Systematics*. Urbana: University of Illinois Press. [Translated book summarizing methods published in earlier papers by Hennig that influenced Brundin.]

Hull, D. 1988. *Science as a Process*. Chicago: University of Chicago Press.

Mayr, E. 1965. "Numerical Phenetics and Taxonomic Theory." *Systematic Zoology* 14: 73–97.

Mead, M. 1928. *Coming of Age in Samoa*. With a foreword by Franz Boas. New York: William Morrow & Co.

Nelson, G. 1973. "Classification as an Expression of Phylogenetic Relationships." *Systematic Zoology* 22: 344–59.

———. 2000. "Ancient Perspectives and Influence in the Theoretical Systematics of a Bold Fisherman." In "Colin Patterson (1933–1998): A Celebration of His Life," ed. P. L. Forey, B. G. Gardiner, and C. J. Humphries. Special issue, *The Linnaean* 2: 9–23.

Nelson, G., and N. Platnick. 1981. *Systematics and Biogeography: Cladistics and Vicariance*. New York: Columbia University Press.

Parenti, L., and M. Ebach. 2009. *Comparative Biogeography. Discovering and Classifying Biogeographical Patterns of a Dynamic Earth*. Berkeley: University of California Press.

Patterson, C. 1964. "A Review of Mesozoic Acanthopterygian Fishes, with Special Reference to Those of the English Chalk." *Philosophical Transactions of the Royal Society of London, Ser. B, Biological Sciences* 247: 213–482.

———. 1978. *Evolution*. London: British Museum (Natural History) and Routledge & Kegan Paul. [2nd ed. published in 1999 by the Natural History Museum, London.]

———, ed. 1987. *Molecules and Morphology in Evolution: Conflict or Compromise?* Cambridge: Cambridge University Press.

Prothero, D. 1999. "Fossil Record." *Encyclopedia of Paleontology*, 490–92. Chicago: Fitzroy Dearborn Publishers.

Raup, D. 1991. *Extinction: Bad Genes or Bad Luck?* New York: W. W. Norton.

Rexova, K., Y. Bastin, and D. Trynta. 2006. "Cladistic Analysis of Bantu Languages: A New Tree Based on Combined Lexical and Grammatical Data." *Naturwissenschaften* 93: 189–94.

Rosen, D., and D. Buth. 1980. "Empirical Evolutionary Research versus Neo-Darwinian Speculation." *Systematic Zoology* 29: 300–308.

Shubin, N. 2008. *Your Inner Fish: A Journey into the 3.5-Billion-Year-Old History of the Human Body.* New York: Pantheon Books.

Simpson, G. G. 1951. *Horses.* New York: Oxford University Press.

———. 1961. *Principles of Animal Taxonomy.* New York: Columbia University Press.

———. 1981. "Exhibit Dismay." *Nature* 290: 286.

Strait, D., F. E. Grine, and M. Moniz. 1997. "A Reappraisal of Early Hominid Phylogeny." *Journal of Human Evolution* 32: 17–82.

Taylor, W. R., and G. C. Van Dyke. 1985. "Revised Procedures for Staining and Clearing Small Fishes and Other Vertebrates for Bone and Cartilage Study." *Cybium* 9, no. 2: 107–19.

Toombs, H. A., and A. E. Rixon. 1959. "The Use of Acids in the Preparation of Vertebrate Fossils." *Curator* 2: 304–12. [Description of the acid-transfer preparation technique.]

Figure Credits

Page xviii: Photo by Lance Grande.

Page 14: Photo by Christopher J. Humphries.

Page 15: Photos by Lance Grande.

Page 16: Photo by Lynne Parenti.

Page 17: Photo by Lance Grande.

Page 18: Image by Lance Grande and Kasey Mennie.

Page 19: Image by Lance Grande and Kasey Mennie.

Page 20: Photo courtesy of Gareth Nelson.

Page 21: Aerial view looking north, taken from a low-flying aircraft in 2014. Image courtesy of Bing Maps, Microsoft Corporation. Thanks to Mark Johnston, who helped me download a high-resolution image from the Bing Maps website.

CHAPTER 2

Notes

1. The main purpose of tenure is to protect free scientific inquiry. A Field Museum board of trustees–approved policy titled "Curatorial Ranks" states: "Free inquiry and expression are essential to the maintenance of research excellence, and career status is essential to free inquiry and expression." Tenure-track curatorial positions are primary research positions, and they make up only a small fraction of the total staff of a natural history museum. Therefore such positions must be awarded with great care. In 2014 only 21 out of 500 Field Museum employees were tenured curators (also see chapter 7, note 1). Sixth-year tenure reviews at the top research institutions are challenging and based primarily on research productivity (mainly publications and grants). Fandos and Pisner (2013) noted that typically fewer than 65 percent of candidates that come up for tenure consideration at Harvard University are successful. A decision to not grant tenure to someone in a tenure-track position prior to their sixth or seventh year means they are to be terminated. By the 1980s, standards for assistant curators in major natural history museums to be promoted to associate curators with tenure became formidable as well, with a significant percentage not making it. When I was hired at the Field Museum, neither of the two previous assistant curators in the geology department had achieved tenure and their positions had been terminated.

Tenure-track ranks for curators, where used, are similar to the ranks used for university professors, with three general levels: assistant curator, associate curator, and curator (sometimes called full curator or senior curator). The vernacular for all three is simply "curator." Most tenure-track curators are hired at the level of assistant curator, which is a position without tenure but with the possibility of it. By their fifth year at the latest, there is an extensive promotion review of a tenure-track assistant curator that has two possible outcomes: the first being promotion to associate curator with tenure, and the second being termination of employment after a set period of a few months to a year. The promotion of an associate curator to full curator usually comes five to twenty-five years after promotion to associate curator and is once again based primarily on research productivity, although other factors are also considered to a lesser degree such as collection-related activity, service, teaching, and administration. There is no set time for promotion at this level, and it is possible for an associate curator to never be promoted. It comes only if a certain level of academic excellence is shown.

In addition to tenure-track curatorial positions, there are also two honorary ones that are generally unsalaried and without tenure but can be renewed as many times as warranted. One is adjunct curator, an appointment usually for professors or other PhD-level scientists who are employed outside the museum but do significant research in the museum. Rarely, non-curatorial scientists with other positions within the museum can also be appointed as adjunct curators if they are consistently active in publishing and other research activities. Another honorary curatorial appointment is curator emeritus, which is given to senior curators of the museum who have retired but continue to come into the museum and do research. The adjunct curators and emeritus curators add to the institution's overall reputation and productivity at little or no cost. In 2013 when there were twenty-one staff curators at the Field Museum, there were also nine emeritus curators and eight adjunct curators.

Another more common honorary research appointment that exists in the museum is the term appointment of research associate, of which there were more than a hundred at the Field Museum in 2014. This appointment is generally made for research professors or curators at other institutions who collaborate with curators at the Field Museum or have some other honorary affiliation that does not generally entail a regular presence in the museum.

2. Over the course of his career, John Bolt (PhD in paleozoology, University of Chicago, 1968) contributed greatly to the museum's rise in stature by investing much of his intellectual energy toward building an exceptional scientific department. This is a vitally important contribution that often goes unrecognized in academia. I certainly owe him personally for giving me my first professional job, but the Field Museum is indebted to him as well for his altruistic vision for the geology department throughout the 1980s. Bolt was curator of vertebrate paleontology at the Field Museum from 1972 to 2008, and continues today as an emeritus curator.

3. Peter R. Crane grew up in Northamptonshire, England, in the blue-collar district of Kettering. As long as I have known him, he has been a globe-trotting, workaholic scientist who thinks big and rarely slows down. His drive and accomplishments have given him license to affect great change in the field of botany and environmental biology. His early collaborative scientific work included some of the first applications of cladistic methodology to early fossil plants. In 1994, together with museum president Sandy Boyd, he created the Field Museum's first major applied environmental conservation program (ECP/ECCo; discussed in chapter 13). In 1999 Peter left the Field Museum to become director and CEO of the Royal Botanic Gardens, Kew, in England, where he took charge of 650 scientists and professional staff along with the world's largest collection of living plants. At Kew, Peter initiated programs of biological inventories around the world and was a driver for one of the largest plant conservation projects in the world: the Millennium Seed Bank Partnership (MSB). The target was to collect seeds of 25,000 plant species that could be banked as insurance against risk of extinction in their native habitat from ever-increasing human activities and climate change. It is a remarkable idea. Once a species is extinct, it is gone forever. But if there are seeds banked, the species can be brought back. The MSB achieved its ten-year target of over a billion seeds. Peter was knighted by the queen of England in 2004 for his work at Kew in horticulture and conservation.

In 2006 Peter received a call from the president of the University of Chicago, offering him a prestigious full professorship. He and wife, Elinor, had already been planning to come back to the states after a ten-year term in England. They moved up their timetable to take advantage of the new job offer. Then in 2010 he left Chicago once again to become dean of the School of Forestry and Environmental Studies at Yale University and curator in charge of paleobotany for the Yale Peabody Museum of Natural History. Over the years, Peter has received four honorary doctorates, he has been elected to the National Academy of Sciences of four different countries, and in 2014 he received the International Prize for biology in Tokyo from the emperor and empress of Japan. In July 2016 Peter became president of a newly formed research center of the Oak Springs Garden Foundation in Virginia. There, he oversees the formation of a major new center for scholarship on a 112-acre site that is devoted to the history and future of plants, gardens, and their roles in society.

4. Scott Lidgard curates the largest of the Field Museum's fossil collections, fossil invertebrates. He is both a paleontologist and marine ecologist who specializes in how body designs evolve in colonial species. A colonial species is one that is composed of multiple constituent organisms, like corals (with many different polyps on a single coral skeleton) or ants (with differently structured workers, drones, and queens). More recently he has been focusing on the history and philosophy of science (e.g., What is a living fossil? How would we identify such a thing? How have our ideas about this changed over time?). Scott is also an award-winning teacher at the University of Chicago. As one of their most popular adjunct professors, he received the 2006 Faculty Award for Excellence in Graduate Teaching and Mentoring.

5. John J. Flynn has done extensive fieldwork around the globe. He has led more than fifty expeditions, including trips to Chile, Madagascar, Mongolia, India, Angola, Colombia, Peru, and the western United States. John is a student and scholar of Earth history, specializing in fossil mammals. Some of his most notable accomplishments include revamping our understanding of carnivore evolution and leading an expedition that found a 20-million-year-old fossil thought to be the oldest well-preserved primate skull from South America. John also coauthored the definitive work delineating the geologic time scale for the Cenozoic (the last 65 million years) used by paleontologists and geologists around the world. In 1998 John became president of the Society of Vertebrate Paleontology, the world's largest organization of specialists studying the evolution of backboned animals. He left the Field Museum in 2004 to become curator and chair of the paleontology division at the American Museum of Natural History in New York. In 2007 he was appointed as the inaugural Dean of the Richard Gilder Graduate School at the American Museum (the first PhD program to be run completely by a natural history museum in the western hemisphere).

6. Olivier Rieppel is the Rowe Family Curator of Evolutionary Biology. Although unpretentious and quiet by nature, he is a maniacally productive sci-

entist and a true intellectual. His office is next to mine, and we are frequently in one another's office discussing science, museum politics, or personal matters. Although brilliant, he has an unfaltering ethical compass and is naturally good-hearted. I think it has been a very therapeutic relationship for the both of us, and he is one of my closest friends at the museum. Much of Olivier's scientific work focuses on the early evolution of reptiles. He has made some amazing paleontological discoveries that fill intuitive gaps in our knowledge about the evolutionary tree of life. In 2008 Olivier and colleagues (Li et al. 2008) discovered a 220-million-year-old turtle from China, *Odontochelys semitestacea*, which turned out to be the oldest known turtle and a transitional form between turtles and other reptiles. It has a bottom shell (plastron) on its belly, but no top shell. Instead, the back has robust ribs, which take little imagination to envision as precursors to a true shell. Another particularly interesting fossil he described with colleagues from Israel (Rieppel et al. 2003 and Tchernov et al. 2000) was a snake with legs, *Haasiophis terrasanctus*, from 93-million-year-old deposits in the Middle East (see page 32). Much of Olivier's better known work lately has been on the general philosophy of science, pondering questions such as "What is a species?" "Are there laws of nature?" "How do we interpret evolutionary process and history from observed patterns in nature?" This is an area where Olivier and I have worked together, and in 1994 we collaborated on a book entitled *Interpreting the Hierarchy of Nature* (Grande and Rieppel 1994).

7. Over the years the five of us have so far collectively produced over 1,400 scientific publications, brought in tens of millions of dollars in grants, added tens of thousands of fossils to the Field Museum collections, curated several highly successful museum exhibits, taught or mentored more than a hundred graduate or postdoctoral students, and become institutional administrators at times, responsible for all the curators and other scientific staff.

References

Berggren, W. A., D. V. Kent, J. J. Flynn, and J. A. Van Couvering. 1985. "Cenozoic Geochronology." *Geological Society of America* 96: 1407–18.

Cameron, M. 2010. "Faculty Tenure in Academe: The Evolution, Benefits, and Implications of an Important Decision." *Journal of Student Affairs at New York University* 10: 1–9.

Crane, P. R. 2013. *Ginkgo: The Tree That Time Forgot*. New Haven, CT: Yale University Press.

Fandos, N., and N. Pisner. 2013. "Joining the Ranks: Demystifying Harvard's Tenure System." *Harvard Crimson*, April 11, 2013.

Grande, L. 1980. *The Paleontology of the Green River Formation with a Review of the Fish Fauna*. Laramie: Geological Survey of Wyoming, Bulletin 63. [2nd ed., 1984.]

Grande, L., and O. Rieppel, eds. 1994. *Interpreting the Hierarchy of Nature: From Systematic Patterns to Evolutionary Process Theories*. San Diego: Academic Press.

Kenrick, P., and P. R. Crane. 1997. *The Origin and Early Diversification of Land Plants: A Cladistic Study*. Washington, DC: Smithsonian Institution Press.

Li, C., X.-C. Wu, O. Rieppel, L.-T. Wang, and L.-J. Zhao. 2008. "An Ancestral Turtle from the Late Triassic of Southwest China." *Nature* 456: 497–501.

McPherson, M. S., and M. O. Schapiro. 1999. "Tenure Issues in Higher Education." *Journal of Economic Perspectives* 13: 85–98.

Padian, K., and K. Angielczyk. 1998. "'Transitional Forms' versus Transitional Features." In *Scientists Confront Intelligent Design and Creationism*, ed. A. Petto and L. Godfrey, 197–230. New York: W.W. Norton.

Rieppel, O., H. Zaher, E. Tchernov, and M. Polcyn. 2003. "The Anatomy and Relationships of *Haasiophis terrasanctus*, a Fossil Snake with Well-Developed Hind Limbs from the Mid-Cretaceous of the Middle East." *Journal of Paleontology* 77: 536–58.

Tchernov, E., O. Rieppel, H. Zaher, M. J. Polcyn, and L. L. Jacobs. 2000. "A Fossil Snake with Limbs." *Science* 287: 2010–12.

Figure Credits

Page 22: Photo by John Weinstein; FM negative #GN88741_53c. Aerial view taken from a helicopter in 2004.

Page 30: Photo by John Weinstein; FM negative #GEO85887_3c.

Page 31: (*Top*) Photo by Diane Alexander White; FM negative #GN87750_34CB; (*bottom left and right*) Images courtesy of Peter Crane.

Page 32: Photo by Mark Widhalm, 1999; FM negative #GN89465_9c.

Page 33: (*Top left*) Photo by John Weinstein, 1997; FM negative #GN88442_2c; (*top right*) Photo courtesy of John Flynn, 1994; FM negative #GEO86061c; (*bottom left*) Photo by John Weinstein, 1997; FM negative #GN88434_7c; (*bottom right*) Photo by John Weinstein, 1996; FM negative #GN88025_8c.

CHAPTER THREE

Note

1. See Grande (2013) for more on this.

Added Commentary

Museum catalog numbers for fossils illustrated in this chapter as well as much more information and history about the Fossil Butte Member are given in the book by Grande (2013) listed below.

References

Grande, L. 1980. *The Paleontology of the Green River Formation with a Review of the Fish Fauna.* Laramie: Geological Survey of Wyoming, Bulletin 63. [2nd ed., 1984.]

———. 2005. "Stones and Bones: Students Learn While Working as Paleontologists." *In the Field* 77: 2–3.

———. 2013. *The Lost World of Fossil Lake: Snapshots from Deep Time.* Chicago: University of Chicago Press.

Julian, J., and the Fossil Country Museum. 2009. *Images of America: Kemmerer.* Charleston, SC: Arcadia Publishing.

Figure Credits

Museum catalog numbers and additional information for fossils illustrated in this chapter are all given in Grande (2013).

Page 34: Photo by Lance Grande.
Page 49: Photo by Lance Grande, 1984.
Page 50: Photos by Lance Grande, 2009.
Page 51: Photos by Lance Grande, 2009.
Page 52: Photos by Lance Grande.
Page 53: Photos by Lance Grande.
Page 54: Photos by Lance Grande and John Weinstein.
Page 55: Photos by Lance Grande and John Weinstein.
Page 56: Photos by Lance Grande and John Weinstein except for top left, which is Mark Mauthner courtesy of Heritage Auctions.
Page 57: Photos by Lance Grande and John Weinstein.
Page 58: Photos by Lance Grande.
Page 59: Photos by Lance Grande.
Page 60: Photos by Lance Grande and John Weinstein.
Page 61: Photos by Lance Grande and John Weinstein.
Page 62: Photo by Lance Grande.
Page 63: Photo by Lance Grande, 2015.

CHAPTER FOUR

Added Commentary

Discussions with Luis Espinosa Arrubarrena were very useful for this chapter. Arrubarrena is a former student and colleague of Shelly Applegate and the current director of the Museum of Geology at the Universidad Nacional Autónoma de México.

References

Applegate, S. P. 1996. "An Overview of the Cretaceous Fishes of the Quarries Near Tepexi de Rodriguez, Puebla, Mexico." In *Mesozoic Fishes*, ed. G. Arratia and G. Viohl, 529–38. Munich: Verlag Friedrich Pfeil.
Butterworth, J. K. 1981. *Latin American Urbanization*. Cambridge: Cambridge University Press.
Grande, L., and W. E. Bemis. 1998. "A Comprehensive Phylogenetic Study of Amiid Fishes (Amiidae) Based on Comparative Skeletal Anatomy: An Empirical Search for Interconnected Patterns of Natural History." *Society of Vertebrate Paleontology Memoir* 4. [*Pachyamia mexicana* from the Tlayúa Formation is described in this book.]
Harner, M. 1977. "The Enigma of Aztec Sacrifice." *Natural History* 86: 46–51.

Figure Credits

Page 64: Photo by Lance Grande.
Page 76: Photo by Pablo Cervantes-Calderon, 1988, sent courtesy of Luis Espinosa Arrubarrena.
Page 77: Photo by Lance Grande.
Page 78: Photo by Lance Grande.
Page 79: Photo by Lance Grande.
Page 80: Photo by Lance Grande.
Page 81: Photos by Lance Grande.
Page 82: Photo by Lance Grande.
Page 83: Photos by Lance Grande.

CHAPTER FIVE

Notes

1. Summarized in Grande (1994) and MacGinitie (1969).
2. The Field Museum, like most major museums and many universities,

use flesh-eating dermestid beetles for the final cleaning of partly defleshed large skeletons. These beetles are harmless to humans because they only eat dead, dried tissue. The dermestids are all kept inside of a special room where there is a series of tanks filled with hundreds of thousands of beetles and their larvae. The partly defleshed skeletons are dried and then placed into the tanks with the insects. The larvae consume muscle and other soft tissue at the rate of several pounds per day, while leaving the bones intact. Once the skeletons are completely free of muscle, skin, and other soft tissues, they are removed from the beetle colony and taken to labs, where they are rinsed, dried, and given catalog numbers keyed to relevant voucher data (e.g., information on where the specimen was collected, its species name, etc.). Then they are ready for incorporation into the museum's main collection. It is amazing that even with all of today's advances in technology, a dermestid beetle colony is still the best way to prepare dry skeletons of medium- to large-size vertebrate animals for study. The dermestid beetle colony room is pressure sealed with a negative pressure inside of a sealed double-door entryway. This secure enclosure is necessary to prevent any beetles from escaping into the rest of the museum where they could do serious damage to collections. The dermestid room is a frequently requested stop for behind-the-scene tour groups but not recommended near lunchtime.

3. Janvire (1998).

4. A few of the major publications that resulted from Willy and my collaborations are listed below (e.g., Grande and Bemis 1991, 1996, 1998, and 1999; and Bemis and Grande 1992, 1999). Some of the material we saw was also used in an 872-page monograph I published on gars (2010). These articles all covered fishes, anatomy, paleontology, biodiversity, and evolution. We rigorously tested computerized methods of cladistics analysis with large amounts of new anatomical data, and some of our publications attracted new students into the field of systematic biology. We also worked together on a comparative anatomy textbook that was widely used by universities around the country (Liem et al. 2001). We were co-advisors for PhD student Eric Hilton, currently a curator and professor at the Virginia Institute of Marine Science. In 2012 I received the Robert H. Gibbs Jr. Memorial Award "for and outstanding body of published work in systematic ichthyology" the highest professional award in systematic ichthyology from North America's largest professional society for ichthyologists and herpetologists, the American Society of Ichthyologists and Herpetologists (see Collett, 2013). My work with Willy played a major role in my receiving that award.

Added Commentary

For most of his career, Willy Bemis (William Elliott Bemis, PhD, University of California, Berkeley, 1983) was a different sort of curator than most of the others mentioned in this book. He didn't start at a well-established museum. His curatorial experience consisted of overseeing a museum-less collection for two decades that was located in the biology department of the University of Massachusetts, Amherst (UMass). The collection included several thousand important zoological and paleontological specimens. For over ten years, Willy added thousands of skeletons to the collection. His operational space included the roof of the university's Morrill Science Center building, where he had a skeletonizing factory powered by a dermestid beetle colony going 24/7.

Willy had a pending request with UMass for a special building to house his planned museum, and for it he amassed a research library with over 3,000 volumes on fishes and raised an endowment of $1.5 million. Only $18 million more and he would have the money needed to endow staff positions for the museum. Pulitzer Prize–winning author John McPhee wrote about Willy's obsessional dream of creating a Massachusetts Museum of Natural History in his critically acclaimed book *The Founding Fish* (2002). In 2005 Willy left UMass and the museum that wasn't quite yet a museum. He accepted a job as director of Cornell University's Shoals Marine Laboratory on Appledore Island, located about seven miles off the coast of Maine. The UMass collection, much of the fish library and the endowment for the museum are still there in Massachusetts waiting for someone else with the credibility, drive, and idealism of Willy to take it further. It is a course of events that demonstrates the value of motivated curatorial researchers to the success of natural history museums.

References

Bemis, W. E., and L. Grande. 1992. "Early Development of the Actinopterygian Head: I. General Observations and Comments on Staging of the Paddlefish *Polyodon spathula*." *Journal of Morphology* 213: 47–83.

———. 1999. "Development of the Median Fins of the North American Paddlefish (*Polyodon spathula*), with Comments on the Lateral Fin-fold Hypothesis." In *Mesozoic Fishes: II. Systematics and the Fossil Record*, ed. G. Arratia and H.-P. Schultze, 41–68. Munich: Verlag Dr. Friedrich Pfeil.

Bemis, W. E., E. J. Hilton, B. Browne, R. Arrindell, A. M. Richmond, C. D. Little, L. Grande, P. L. Forey, and G. J. Nelson. 2004. "Methods for Preparing Dry,

Partially Articulated Skeletons of Osteichthyans, with Notes on Making Ridewood Dissections of the Cranial Skeleton." *Copeia*, no. 3: 603-9.

Collette, B. 2013. "Lance Grande: Robert H. Gibbs, Jr. Memorial Award for Excellence in Systematic Ichthyology—2012." *Copeia*, no. 1 (2013): 2-3.

Grande, L. 1994. "Repeating Patterns in Nature, and 'Impact' in Science." In *Interpreting the Hierarchy of Nature: From Systematic Patterns to Evolutionary Process Theories*, ed. L. Grande and O. Rieppel, 61-84. San Diego: Academic Press.

——. 2010. *An Empirical Synthetic Pattern Study of Gars (Lepisosteiformes) and Closely Related Species, Based Mostly on Skeletal Anatomy: The Resurrection of Holostei*. Special publication, American Society of Ichthyologists and Herpetologists. [Description of *Masillosteus janaea* on pp. 635-61.]

Grande, L., and W. Bemis. 1991. "Osteology and Phylogenetic Relationships of Fossil and Recent Paddlefishes (Polyodontidae) with Comments on the Interrelationships of Acipenseriformes." *Society of Vertebrate Paleontology Memoir* 1. [Willy Bemis and my first monograph, with extensive use of combination figures illustrating rayfin skeletons, an illustration technique that we further developed in Grande and Bemis (1998); Grande (2010), and Hilton, Grande, and Bemis (2011).]

——. 1996. "Interrelationships of Acipenseriformes, with Comments on 'Chondrostei.'" In *Interrelationships of Fishes*, ed. M. Stiassny, L. Parenti, and D. Johnson, 85-115. San Diego: Academic Press.

——. 1998. "A Comprehensive Phylogenetic Study of Amiid Fishes (Amiidae) Based on Comparative Skeletal Anatomy: An Empirical Search for Interconnected Patterns of Natural History." *Society of Vertebrate Paleontology Memoir* 4.

——. 1999. "Historical Biogeography and Historical Paleoecology of Amiidae and Other Halecomorph Fishes." In *Mesozoic Fishes: II. Systematics and the Fossil Record*, ed. G. Arratia and H.-P. Schultze, 413-24. Munich: Verlag Friedrich Pfeil.

Hilton, E., L. Grande, and W. E. Bemis. 2011. "Skeletal Anatomy of the Shortnose Sturgeon, *Acipenser brevirostrum* Lesueur, 1818, and the Systematics of Sturgeons (Acipenseriformes, Acipenseridae)." *Fieldiana, Life and Earth Sciences*, no. 3: 1-168.

Janvire, P. 1998. "Bowfins and the Revenge of Comparative Biology." *Science* 281: 1150.

Liem, K. F., W. E. Bemis, W. F. Walker, and L. Grande. 2001. *Functional Anatomy of the Vertebrates: An Evolutionary Perspective*, 3rd ed. Fort Worth: Saunders College Publishers.

MacGinitie, H. D. 1969. "The Eocene Green River Flora of Northwestern Colorado and Northeastern Utah." *University of California Publications in the Geological Sciences* 83.

McPhee, J. 1998. "Catch-and-Dissect." *New Yorker*, October 19, 58–66.

———. 2002. *The Founding Fish*. New York: Farrar, Straus and Giroux.

Nur, A., and Z. Ben-Avraham. 1977. "Lost Pacifica Continent." *Nature* 270: 41–43.

Paddock, R. 1997. "Central Heat Saps Moscow's Economy." *Los Angeles Times*, March 23.

Simpson, G. G. 1943. "Mammals and the Nature of Continents." *American Journal of Science* 241: 1–31. [One of Simpson's early arguments against continental drift.]

———. 1965. *The Geography of Evolution*. New York: Capricorn Books. [Simpson still rejecting the notion of drifting continents.]

Figure Credits

Page 84: Photo by Willy Bemis, 2002.

Page 93: Image by Lance Grande and Kasey Mennie. After Nur and Ben-Avraham (1977).

Page 102: Photos by Lance Grande.

Page 103: Photo by Lance Grande.

Page 104: Photo by Lance Grande.

Page 105: Photo by Lance Grande.

Page 106: Photo by Willy Bemis.

Page 107: Photos by Lance Grande.

Page 108: Photo by Steve Droter © Chesapeake Bay Program, 2013.

Page 109: Photo and line drawing by Lance Grande (from Grande 2010).

CHAPTER SIX

Notes

1. Robert T. Bakker, PhD in paleontology from Harvard University, 1976. Bakker is portrayed in Steven Spielberg's *The Lost World of Jurassic Park* as the bearded paleontologist Robert Burke, who ends up in the movie getting eaten by a *Tyrannosaurus rex*.

2. The Society of Vertebrate Paleontology, founded in 1940, is a scientific organization with about 2,000 members whose stated mission is to advance the science of vertebrate paleontology. It serves as the largest organization of professional vertebrate paleontologists but is also open to anyone interested in vertebrate paleontology. Part of its mission is the conservation and preservation of important vertebrate fossil sites, particularly those on public lands.

3. The original indictment of 39 counts against the BHI and various individuals within the company was filed in the U.S. District Court for the District of South Dakota Western Division on November 19, 1993, for (1) conspiracy, (2)

entry of goods by means of false statement, (3) theft of government property, (4) false statements, (5) wire fraud, (6) obstruction of justice, (7) money laundering, (8) interstate transportation of stolen goods, (9) structuring, and (10) currency or monetary instrument report violation.

4. Part of the evidence used in the trial against the BHI was a collection of fossil catfishes from the South Dakota Badlands. I had purchased these from them for the Field Museum, and they had been labeled as coming from a legal collecting site. Later it was found that the locality data we were given was incorrect and that they were actually from a National Park area. At the end of the trial, the rest of the fossil catfishes from that site that had been confiscated from the BHI during the federal raid were returned to the National Park Service, who subsequently transferred them to the Field Museum on long-term loan. I hope to have the rest of these prepared in the next few years so I can publish a scientific description of the material.

5. Christopher A. Brochu, PhD in geological sciences from the University of Texas, Austin, 1997.

6. Peter J. Makovicky, PhD in Earth and environmental sciences from Columbia University, 2001. Pete was a student in the vertebrate paleontology department of the American Museum, which had its collaborative PhD program with Columbia University. I was a student in the ichthyology department of the museum, which had its collaborative PhD program with the City University of New York.

7. The SUE trials generated many articles in law journals around the country, only a few of which are listed here.

Added Commentary

Verbal or e-mail discussions in 2014 with Peter Larson, Vince Santucci, Nate Eimer, Sandy Boyd, John McCarter, Bill Simpson, and Peter Makovicky were useful in the writing of this chapter. Some of the quotes and details were taken from Fiffer (2000) or Larson and Donnan (2002).

References

Brochu, C. A. 2003. "Osteology of *Tyrannosaurus rex*: Insights from a Nearly Complete Skeleton and High-Resolution Computed Tomographic Analysis of the Skull." *Society of Vertebrate Paleontology Memoir 7*. [Illustrated description of the skeleton of SUE.]

Duffy, P. K., and L. A. Lofgren. 1994. "Jurassic Farce: A Critical Analysis of the Government's Seizure of 'Sue,' a Sixty-Five-Million-Year-Old *Tyrannosau-*

rus rex Fossil." *South Dakota Law Review* 39: 47528. [Duffy was Pete Larson's lawyer.]

Dussias, A. M. 1996. "Science, Sovereignty, and the Sacred Text: Paleontological Resources and Native American Rights." *Maryland Law Review* 55: 84–159.

Fiffer, S. 2000. *Tyrannosaurus Sue: The Extraordinary Saga of the Largest, Most Fought Over* T. rex *Ever Found*. New York: W. H. Freeman.

Grande, L. 1980. *The Paleontology of the Green River Formation with a Review of the Fish Fauna*. Laramie: Geological Survey of Wyoming, Bulletin 63. [2nd ed., 1984.]

Hutchinson, J. R., K. T. Bates, J. Molnar, V. Allen, and P. J. Makovicky. 2011. "A Computational Analysis of Limb and Body Dimensions in *Tyrannosaurus rex* with Implications for Locomotion, Ontogeny, and Growth." *PLoS ONE* 6, no. 10: e26037. DOI: 10.1371/journal.pone.0026037. [Weight, growth and age estimates for SUE.]

Larson, P., and K. Donnan. 2002. *Rex Appeal: The Amazing Story of Sue, the Dinosaur That Changed Science, the Law, and My Life*. Montpelier, VT: Invisible Cities Press. [Donnan is Pete Larson's ex-wife.]

Lazerwitz, D. J. 1994. "Bones of Contention: The Regulation of Paleontological Resources on the Federal Public Lands." *Indiana Law Journal* 69: 601–36.

Poindexter, M. D. 1994. "Of Dinosaurs and Indefinite Land Trusts: A Review of Individual American Indian Property Rights Amidst the Legacy of Allotment." *Boston College Third World Law Journal* 14: 53–81.

Sink, M. 1999. "In the Earth's Graveyard, Little Protection against Theft." *New York Times*, June 15. [Vince Santucci is described as "the only pistol-packing paleontologist of the National Park Service."]

Figure Credits

Page 110: Photo by Mark Widhalm; Field Museum negative #GN89775_12c. Photo taken in 2000, late one night after the museum closed.

Page 141: Photo by Peter L. Larson; © 1990 Black Hills Institute of Geological Research.

Page 142: Photo by Susan Hendrickson; © 1990 Black Hills Institute of Geological Research.

Page 143: Photo © 1990 Black Hills Institute of Geological Research; photographer: Peter L. Larson (*top*), and bottom image redrawn after image copyrighted by 1990 Black Hills Institute of Geological Research by Peter L. Larson and Larry Schaffer.

Page 144: Photo by Peter L. Larson; © 1990 Black Hills Institute of Geological Research.

Page 145: Photo by Louie Psihoyos; Corbis image #42-21814815.

Page 146: Photos © 1992. Courtesy of Black Hills Institute of Geological Research.

Page 147: Photo courtesy of David Redden.

Page 148: Photo by John Weinstein; Field Museum negative #GN89035_28c.
Page 149: Photo by John Weinstein; Field Museum negative #GN89145_5CB.
Page 150: Photo by John Weinstein, 2000; Field Museum negative #GN89656_22CB.
Page 151: Photos by John Weinstein; Field Museum negatives #GN89693_25Ac (*top*) and GN89752_2Ac (*bottom*).
Page 152: Photo by John Weinstein; Field Museum negative #GN89709_32Ac.
Page 153: Photo by John Weinstein, March 2014; Field Museum negative #GN91961_073Cd.
Page 154: Images of scans made by the Forensic Services Division of the Chicago Police Department. Courtesy of Pete Makovicky.
Page 155: Photo by Lance Grande, April 2014.

CHAPTER SEVEN

Notes

1. One question that is inevitably debated within an institution is what is the ideal number of curators? The answer depends on many factors including the composition of the collection, the availability of world-class specialists during curatorial hiring searches, and institutional budget constraints. The number of curatorial research positions in any museum is generally a small fraction of the total full-time staff. Determining the ideal proportion of curatorial positions to other museum positions can be a challenge for any institution. In 2015 at the big three U.S. museums, the Smithsonian National Museum of Natural History had 81 curatorial positions out of an estimated total staff of 1,100 (8.1%), the American Museum of Natural History had 41 curators out of a total staff of 1,012 (4.1%), and the Field Museum had 21 curatorial positions out of an estimated total staff of 500 (4.2%). Some of the variance in total staff numbers among the museums is due to differential use of non-staff contractors for certain positions (e.g., security, stores, and restaurants), which affects the way the number of full-time staff is counted. Over the last 120 years, the number of curatorial positions has oscillated from 28 to 124 at the National Museum (peak number in the 1980s), 11 to 52 at the American Museum (peak number in the 1960s), and 7 to 38 at the Field Museum (peak number in the early 2000s). The National Museum has a federally supported operating budget, giving them the greatest capacity to support research curators. Other natural history museums must rely on their own revenue-generating operations and donations for sustainability. Recent reductions in numbers of curators at the three institutions correspond to reallocation of scientific funds due to broadened programs (e.g., applied conservation programs, new postdoctoral and educational positions) and other budget chal-

lenges, opportunities, and constraints. Numbers of curatorial positions over time at the Field Museum and the American Museum were taken from online annual reports. Numbers for the Smithsonian National Museum of Natural History were provided courtesy of Kirk Johnson and Mike McCarthy of the NMNH.

2. Gregory M. Mueller, PhD in botany, University of Tennessee, Knoxville, 1982. Greg was the Field Museum's curator of mushrooms and fungi (mycology) from 1985 to 2009. Today Greg is the Negaunee Foundation Vice President of Science at the Chicago Botanic Garden.

3. Richard H. Ree, PhD in organismic and evolutionary biology, Harvard University, 2001.

4. H. Thorsten Lumbsch, PhD in botany, University of Essen, Germany, 1993.

5. Michael O. Dillon, PhD in botany, University of Texas, Austin, 1976.

6. John M. Bates, PhD in zoology, Louisiana State University, 1993.

7. Today, after 120 years of growth, the bird collection of the Field Museum contains more than a half million specimens. Of those there are about 400,000 stuffed skins, 80,000 skeletons, 22,000 egg sets (a set being a nest clutch or other associated collection), 20,000 bird bodies preserved in alcohol, and 85,000 frozen tissue samples for DNA analysis. Another major bird salvage operation I saw at the museum in 2005 (in addition to the annual collecting of dead birds that have crashed into buildings in the city of Chicago) was a collection of over 600 large great grey owls that had been killed over a short period of time by cars and trucks on a highway in northern Minnesota. The natural range of the species had shifted south due to severe winter conditions in Canada, and the unfamiliar hazards turned out to be disastrous for the owls. Today we are involved in many types of salvage programs at the Field Museum. In fact, salvage is now by far the largest source of collection growth for the Bird Division.

8. Meenakshi Wadhwa, PhD in Earth and planetary sciences from Washington University, St. Louis, 1994.

9. Philipp R. Heck, PhD in cosmochemistry from the Swiss Federal Institute of Technology in Zurich, Switzerland, 2005.

10. Kenneth D. Angielczyk, PhD in integrative biology, University of California, Berkeley, 2003.

11. Janet R. Voight, PhD in ecology and evolutionary biology from the University of Arizona, 1990.

12. Petra Sierwald, PhD in biology from the University of Hamburg, 1985.

13. Shannon J. Hackett, PhD in evolutionary biology from Louisiana State University, 1993.

14. Corrie S. Moreau, PhD in organismic and evolutionary biology from Harvard University, 2007.

15. There are significantly fewer women than men in the curatorial profession, even though some of the profession's most famous curators have been women, such as Margaret Mead. As of June 2014, I surveyed the big three natural

history museums of North America based on published annual reports and came up with the following figures. At the Smithsonian National Museum of Natural History, 30 percent of the curators are women; at the American Museum of Natural History, 20 percent are women; and at the Field Museum, 20 percent are women. The proportions of women curators at most other major natural history museums are similarly low, including 25 percent at the Harvard Museum of Comparative Zoology, 14 percent at the Yale Peabody Museum of Natural History, and 7 percent at the California Academy of Sciences. As low as these numbers are today, they represent improvements. While I was at the American Museum from 1979 to 1983, the percentage of staff curators who were women was less than 10 percent; and there were no women curators on staff when I started at the Field Museum in 1983.

Why the lopsided ratios? There are several reasons. Recent studies show that there is still some degree of gender bias when it comes to hiring in academic institutions, even by academics themselves (Moss-Racusin et al. 2012). Although gender bias in hiring may still be a factor for the gender discrepancy, I believe that a larger reason is the small pool of qualified women applicants to begin with, particularly in systematics and paleontology. I can speak from personal experience here. The last curatorial search that I chaired in the Field Museum's department of geology was in 1989 for a curator of vertebrate paleontology. The search resulted in 111 applications from people with PhDs, of which only 7 applicants (6%) were women. In every one of the eleven curatorial searches I have been involved in over the last twenty-eight years, the male applicants have far outnumbered the female ones. Recent studies of women in science by the American Association of University Women have indicated that qualified women who apply for positions at top research institutions are more likely to be hired than men, but smaller percentages of qualified women apply for such positions in the first place (Hill, Corbett, and St. Rose 2010). The problem is compounded by the competitive bidding among academic institutions for existing top women scientists. While I was head of the Field Museum's Collections and Research division, one woman curator (Meenakshi Wadhwa in this chapter) was lured away to Arizona State University with an offer so huge, that it would have required cannibalizing several curatorial positions to match, even if those positions had been available.

One reason for the low proportion of women candidates for curatorial positions is the lack of support and encouragement for young girls to pursue interests in science early in their educational trajectories (Ceci and Williams 2010). I suspect that this is especially true for fields such as comparative anatomy and systematic biology, two fields of particular importance for natural history museum curators. Also, many young girls who initially become interested in the field of biology may currently be channeled into other directions such as medicine during their college and university training (the percentage of women in medical school has risen sharply during the last forty years). Studies also indicate that there are some women who choose not to pursue certain scientific fields because of the coincidence of childbearing years with the period of time when academic researchers must build a strong research portfolio in order to get tenure (McNutt 2013).

I cannot provide a definitive solution to the overall challenges here, but it seems the best way to address the scarcity of top women scientists in the curatorial profession is to start in grammar school and high school to encourage and nurture the early interest of girls who might wish to enter professions such as systematic biology, evolutionary biology, or paleontology. Postponement of tenure evaluation deadlines to accommodate childbirth could also be given consideration and made clear in posted job descriptions for curators. And programs such as Corrie Moreau's Women in Science group provide much-needed encouragement for women at an early stage in their aspirations to become curators as well as other research scientists.

16. Gary M. Feinman, PhD in anthropology, City University of New York, 1980.

17. P. Ryan Williams, PhD in anthropology from the University of Florida, 1997.

18. William A. Parkinson, PhD in anthropology from the University of Michigan, 1999.

19. John E. Terrell, PhD in anthropology from Harvard University, 1976.

20. Robert D. Martin, PhD in animal behavior from Worcester College, Oxford, England, 1967. Prior to coming to the Field Museum in 2001, Bob was director and professor of the Anthropological Institute in Zurich, Switzerland. From 2001 to 2006, Bob was vice president and provost for the Field Museum, after which he left administration to become curator of physical anthropology. Bob retired in 2013 and became curator emeritus at the museum, and today he splits his time between Chicago and Europe (where he continues to teach primate evolution in Paris through the University of Chicago Study Abroad program).

21. Chapurukha M. Kusimba, PhD in anthropology from Bryn Mawr College, 1993. He was the only African in his class and graduated as the class valedictorian. He joined the Field Museum's curatorial staff on July 1, 1994. In 2013 Chap left the Field Museum to become professor and chair of the anthropology department at the American University in Washington, DC, where he continues to actively pursue his research.

References

Bates, J. M. 2007. "Natural History Museums: World Centers of Biodiversity Knowledge, Now and in the Future." *Systematist* 29: 3–6.

Bates, J., and T. C. Demos. 2001. "Do We Need to Devalue Amazonian and Other Large Tropical Forests?" *Diversity and Distribution* 7: 249–55.

Ceci, S. J., and W. Williams, 2010. "Sex Differences in Math-Intensive Fields." *Current Direction in Psychological Science* 19: 275–79.

Diehl, R., J. Bates, D. Willard, and T. Gnoske. 2014. "Bird Mortality during Migration over Lake Michigan: A Case Study." *Wilson Journal of Ornithology* 126: 19–29.

Eaton, D., and R. Ree. 2013. "Inferring Phylogeny and Introgression Using RADseq Data: An Example from Flowering Plants (*Pedicularis*: Orobanchaceae)." *Systematic Biology* 62: 689–706.

Elkington, B. G., K. Sydara, A. Newsome, C. H. Hwang, D. C. Lankin, C. Simmler, J. G. Napolitano et al. 2014. "New Finding of an Anti-TB Compound in the Genus *Marsypopetalum* (Annonaceae) from a Traditional Herbal Remedy of Laos." Journal of Ethnopharmacology 151, no. 2: 903–11.

Gould, S. J. 1977. *Ever Since Darwin: Reflections in Natural History*. New York: W. W. Norton.

Hackett, S. J., R. T. Kimball, S. Reddy, R. C. K. Bowie, E. L. Braun, M. J. Braun, J. L. Chojnowski et al. 2008. "A Phylogenomic Study of Birds Reveals Their Evolutionary History." *Science* 320: 1763–68.

Hamill, S. 2004. "Park Forest Meteorite Fall: One Year Many Deals Later." *Chicago Tribune*, March 26.

Heck, P., B. Schmitz, H. Baur, A. Halliday, and R. Wieler. 2004. "Fast Delivery of Meteorites to Earth after a Major Asteroid Collision." *Nature* 430: 323–25.

Hill, C., C. Corbett, and A. St. Rose. 2010. *Why So Few? Women in Science, Technology, Engineering, and Mathematics*. Washington, DC: AAUW.

Kammerer, C., K. Angielczyk, and J. Fröbisch. 2011. "A Comprehensive Taxonomic Revision of *Dicynodon* (Therapsida, Anomodontia), and Its Implications for Dicynodont Phylogeny, Biogeography, and Biostratigraphy." *Society of Vertebrate Paleontology Memoir* 11: 1–158.

Korochantseva, E., M. Trieloff, C. Lorenz, A. Buykin, M. Ivanova, W. Schwarz, E. Jessberger. 2007. "L-chondrite Asteroid Breakup Tied to Ordovician Meteorite Shower by Multiple Isochron 40Ar-39Ar Dating." *Meteoritics and Planetary Science* 42: 113–30.

Kusimba, C. 1997. "A Time Traveler in Kenya." *Natural History* 106: 38–47.

Lumbsch, H. T. 2009. "Not a Drop to Drink: How Plants Survive Waterless Conditions by Sticking Together." *In the Field*.

Lumbsch, H. T., P. M. McCarthy, and W. M. Malcolm. 2001. "Key to the Genera of Australian Lichens: Apothecial Crusts." Canberra, AUS: ABRS Government Printer.

Martin, R. D. 2013. *How We Do It: The Evolution and Future of Human Reproduction*. New York: Basic Books.

Martin, R. D., A. M. MacLarnon, J. L. Phillips, L. Dussubieux, P. R. Williams, and W. B. Dobyns. 2006. "Comment on 'The Brain of LB1, *Homo floresiensis*.'" *Science* 312: 999.

McFarland, J., and G. Mueller. 2009. *Edible Wild Mushrooms of Illinois and Surrounding States: A Field to Kitchen Guide.* Urbana: University of Illinois Press.

McNutt, M. 2013. "Leveling the Playing Field." *Science* 341: 317.

Moreau, C. S., C. D. Bell, R. Vila, S. B. Archibald, and N. P. Pierce. 2006. "Phylogeny of the Ants: Diversification in the Age of Angiosperms." *Science* 312: 101–4.

Moseley, M., D. Nash, P. Williams, S. DeFrance, A. Miranda, and M. Ruales. 2005. "Burning Down the Brewery: Excavation and Evacuation of an Ancient Imperial Colony at Cerro Baúl, Perú." *Proceedings of the National Academy of Sciences* 102: 17264–71.

Moss-Racusin, C. A., J. F. Dovidio, V. L. Brescoll, M. J. Graham, and J. Handelsman. 2012. "Science Faculty's Subtle Gender Biases Favor Male Students." *Proceedings of the National Academy of Sciences* 109: 16474–79.

Price, T., and G. Feinman. 2013. *Images of the Past.* 7th ed. New York: McGraw-Hill.

Schmitz, B., M. Lindstrom, F. Asaro, and M. Tassinari. 1996. "Geochemistry of Meteor-Rich Limestone Strata and Fossil Meteorites from the Lower Ordovician at Kinnekulle, Sweden." *Earth and Planetary Letters* 145: 31–48.

Terrell, J. 2014. *A Talent for Friendship: Rediscovery of a Remarkable Trait.* Oxford: Oxford University Press.

Wilson, E. O., 2014. *Letters to a Young Scientist.* New York: Liveright.

Winger, B., F. Barker, and R. Ree. 2014. "Temperate Origins of Long-Distance Seasonal Migration in New World Songbirds." *PNAS* 111: 12115–20.

Winker, K., J. Reed, P. Escalante, R. Askins, C. Cicero, G. Hough, and J. Bates. 2010. "The Importance, Effects, and Ethics of Bird Collecting." *The Auk* 127: 690–95.

Figure Credits

Page 156: Image by Lance Grande and Kasey Mennie.

Page 180: (*Top*) Photo by Betty Strack, 2003; (*bottom left*) Photo by Susan Kelly, 2007; (*bottom right*) Photo by Khwanruan Papong, 2011.

Page 181: Photo by Lance Grande.

Page 182: Photo by Lance Grande.

Page 183: (*Top left*) Photo by Sanchez Vega courtesy of Michael Dillon; (*top right*) Photo by Lance Grande; (*bottom*) Image from IMOD website.

Page 184: Photo courtesy of Mike Dillon; photographer James Hendrickson.

Page 185: (*Top left*) Photo by Josh Engle; (*top right and bottom*) Photos by Lance Grande.

Page 186: (*Top left*) Photo by Mark Widhalm, Field Museum negative #GN90414_07d, holding meteorite specimen number ME340; (*top right*) Photo by Lance Grande, holding meteorite specimen number ME6048.1; (*bottom*) Photo courtesy of Meenakshi Wadhwa.

Page 187: (*Top left*) Photo by John Weinstein, Field Museum negative #GN91375_22Ad; (*top right*) Photo by Lance Grande; (*bottom left*) Photo courtesy of Ken Angielczyk, 2006; (*bottom right*) Image modified after a drawing of Marlene Donnelly for Ken Angielczyk.

Page 188: Photos courtesy of Janet Voight and Woods Hole Oceanographic Institute, 2003.

Page 189: (*Top left*) Photo by Karen Bean; Field Museum negative #GN91313_087d; (*top right*) Photo by John Weinstein; Field Museum negative #GN90809_37d; (*bottom*) Photo by John Weinstein; Field Museum negative #GN91147_013Ad.

Page 190: Photographer Roberto Keller Perez, 2014. Photo courtesy of Corrie Moreau.

Page 191: (*Top left*) Photo by Linda Nicholas, 2014; (*top right*) Photo Linda Nicholas, 2005; (*bottom left and right*) Photos courtesy of Ryan Williams and the Cerro Baúl Archaeological Project.

Page 192: Photos courtesy of Bill Parkinson.

Page 193: Photo courtesy of Bill Parkinson.

Page 194: (*Top left*) Photo by Sampson Purupuru and courtesy of John Terrell; (*top right*) Photo by Lance Grande; (*bottom left*) Photo by John Weinstein, 2007; Field Museum negative #GN91016_45d; (*bottom right*) Photo by Lance Grande.

Page 195: Photo by John Weinstein, 1998; Field Museum negative #Z987306c.

CHAPTER EIGHT

Notes

1. Today selective collecting of animals is done in the most humane way possible following very strict guidelines. Endangered species are not taken, although specimens of many endangered species are already available in museums because they were collected when the species were more abundant or before they had been identified as new species. As Bruce Patterson points out in his 2002 article "On the Continuing Need for Scientific Collecting of Mammals," there is no known case where scientific collecting of animals has contributed to the peril or extinction of an animal species. On the contrary, information from these collections ultimately helps *promote* the survival of the species and ecosystems from where these collections originally came. Preserved specimens are critical for study of an animal's genetics, parasites, internal anatomy, stomach contents, ecology, evolutionary relationships, and other features. These date-specific collections document baseline aspects of biology and ecology for comparison with past and future samples. In this way we can analyze the ongoing health of animal

and plant communities in the face of ecological disasters and global climate change.

Natural history collections often help solve problems that were not envisioned at the time of their acquisition. Examples at the Field Museum include DNA studies being done today on specimens that were collected over a century ago to trace the origins of disease or changing ecosystems. One of my favorite examples of an unanticipated use of a collection was in the late twentieth century when studies of the eggshell collection of the museum led to saving the bald eagle from extinction (summarized in Wurster 2015). Eagles were at one time highly endangered with populations plummeting to dangerously low levels. Studies of eagle eggs collected over the years going back a century indicated that the falling population of eagles was the result of their eggshells becoming extremely thin. The time when the general eggshell thinning in eagles began was correlated with the time DDT started to be widely used in agriculture. This finding eventually resulted in restrictions in the use of DDT, which was followed by resurgence in the bald eagle population.

Having a collection of cataloged specimens that are preserved in perpetuity by museums also makes it unnecessary for scientists to kill specimens for many future studies requiring physical examination of animals. When I did my doctoral research on the evolution and comparative anatomy of clupeomorph fishes including the herring, anchovies, and other herring-like species (published as Grande 1985), I did not have to make new collections of any live animals. The alcohol-preserved collections of the American Museum, the Field Museum, the Smithsonian National Museum of Natural History, and the London Natural History Museum gave me access to 155 different species of clupeomorph fishes from which I was able to make cleared and stained skeletal preparations for analysis.

References

Grande, L. 1985. "Recent and Fossil Clupeomorph Fishes with Materials for Revision of the Subgroups of Clupeoids." *Bulletin of the American Museum of Natural History* 181: 231–372.

Patterson, B. D. 2002. "On the Continuing Need for Scientific Collecting of Mammals." *Mastozoologia Neotropical: Journal of Neotropical Mammalogy* 9: 253–62.

Pope, C. H. 1958. "Fatal Bite of Captive African Rear-Fanged Snake (*Dispholidus typus*)." *Copeia* no. 4: 280–82.

Schmidt, K. P. 1923 and 1957. [Schmidt's field notes from the Field Museum's archives.]

———. 1952. *Crocodile Hunting in Central America*. Popular Series: Zoology, no. 15. Chicago: Chicago Natural History Museum.

Wright, A. G. 1967. *In the Steps of the Great American Herpetologist, Karl Patterson Schmidt*. New York: M. Evans Company.

Wurster, C. 2015. *DDT Wars: Rescuing Our National Bird, Preventing Cancer, and Creating EDF*. New York: Oxford University Press.

Figure Credits

Page 196: Photographer unknown; Field Museum negative #Z86242 taken in 1952. The lizards are helmeted iguanas (*Corytophanes cristatus*) from Gallon Jug, Belize.

Page 204: Photographer unknown, 1923; Field Museum negative #CSZ47726.

Page 205: Photographer unknown; Field Museum negative # CSZ47695 (*top*) and CSZ47699 (*bottom*).

Page 206: Photos by Lance Grande.

Page 207: Photographer unknown; Field Museum negative #GN91736_478c.

Page 208: Photos by Johan Morais.

Page 209: Photo by Lance Grande.

CHAPTER NINE

Notes

1. The eight vice presidents during most of the years I headed the Collections and Research division of the museum included the vice president for Collections and Research (my position), the vice president for Environment, Culture, and Conservation (ECCo), the vice president for Technology, the vice president for Institutional Advancement (fund-raising), the vice president for Administration, the vice president for Exhibitions and Museum enterprise, the vice president for Board Relations, and the vice president for Operations.

2. The mission statement of the Field Museum has remained remarkably steady over the years. In September 1893, the founders of the museum dedicated the institution to "the accumulation and dissemination of knowledge." From the start, its research focus has been in the areas of anthropology, biology, and geology. On June 15, 1992, the museum's board of trustees approved a new mission statement including: "The Field Museum is an educational institution concerned with the diversity and relationships in nature and among cultures. It provides collection-based research and learning for greater public understanding and appreciation of the world in which we live." In this document the research focus was specified as being centered on "anthropology and the natural sciences of evolutionary and environmental biology and geology." Then in a document distributed to the board of trustees on September 27, 1999, the mission statement included: "We rededicate the Museum for a new century to the creation, accumulation and dissemination of knowledge," nearly identical to the mission statement of a century earlier. Finally, in 2014 the Field Museum lists the mission statement on its website: "The Field Museum inspires curiosity about

life on Earth while exploring how the world came to be and how we can make it a better place. We invite visitors, students, educators and scientists from around the world on a journey of scientific discovery. Our exhibits tell the story of life on Earth. Our collections solve scientific mysteries. Our research opens new vistas. Our science translates into action for a healthy planet. As educators, we inspire wonder and understanding." The stated research focus today remains the same but uses different divisional names (social sciences instead of anthropology, life sciences instead of biology, and earth sciences instead of geology).

3. The functional heart of the *Encyclopedia of Life* was a series of interconnected websites—with the ultimate goal of having a page for each of the nearly 2 million species on the planet. Each digital page is designed to be infinitely expandable, including sound, video, still images, graphics, text, and links to other web content from any country. The number of digital species pages continues to expand indefinitely as species and different varieties of content are added and organized by scientists around the world who function as content gatekeepers. Existing websites and online data files are incorporated into the *EOL*, and new content is generated by biodiversity experts, enthusiastic amateurs, and just about anybody with something significant to add, using the citizen-science approach to expanding the website. In this way, thousands of people around the world are contributors to the ever-growing database. Mark Westneat (curator of fishes at the Field Museum from 1992 to 2013) was the director of Field Museum's functional unit of *EOL*, the Biodiversity Synthesis Center (BioSynC). The main job of BioSynC was to jump-start participation in *EOL* by the world's greatest biodiversity specialists. Over the years it hosted over forty major symposia including over a thousand visiting scientists from forty-five countries under Mark's excellent directorship. Even after initial grant funding ended for this phase of *EOL*, the BioSynC continued for another two years through additional grants obtained by Mark and the BioSynC staff. In 2013 the BioSynC division of the *EOL* came to a close after Mark moved to the University of Chicago to become professor of organismal biology and anatomy.

4. There has been massive destruction of museums and antiquities in Iraq, Afghanistan, and Syria over the last decade by Islamic extremists who view the artifacts as idols offensive to their religious beliefs. Numerous articles in the press have been documenting this regularly (e.g., Bernard 2001; Cambanis 2015; Cullinane, Alkhshali, and Tawfeeq 2015; Shaheen 2015). As is often the case through human history, religious extremism periodically clashes with the mission of science and cultural historians, eroding empirical evidence and knowledge of our history forever.

5. Support from the University of Chicago for the Field Museum's Collection and Research division during the peak of the museum's fiscal crisis was facilitated by some of my colleagues at the university to whom I will always feel a debt of gratitude. These include Professor Mike Foote (chair of Geophysical Sciences),

Professor Joy Bergelson (chair of Ecology and Evolution), Michael Coats (chair of the Committee on Evolutionary Biology), and especially my colleague Neil Shubin (associate dean for Academic Strategy). Also, the university's assistance would not have been possible without the support of University of Chicago provost Tom Rosenbaum (now president of California Institute of Technology).

Added Commentary

Doing the job of heading the Collections & Research division of the Field Museum would not have been possible for me without the amazing office staff, including Carter O'Brien, Mark Alvey, and Towanda Simmons, who was later replaced with Kasey Mennie. The mountain of organizational responsibility could be challenging at times, but this group made my job possible, and I am glad to be able to thank them here.

References

Atkinson, R. C., and W. A. Blanpied. 2007. "Research Universities: Core of the U.S. Science and Technology System." *Technology and Society* 30: 30-48.

Bernard, P. 2001. "What the Taliban Destroyed." *Wall Street Journal*, December 20.

Bonney, R., H. Ballard, R. Jordan et al. 2009. *Public Participation in Scientific Research: Defining the Field and Assessing Its Potential for Informal Science Education*. A CAISE Inquiry Group Report. Washington, DC: Center for Advancement of Informal Science Education (CAISE). [A paper on the citizen-science approach to scientific research.]

Cambanis, T. 2015. "Why ISIS' Destruction of Antiquities Hurts So Much." *Boston Globe*, March 10.

Cole, J. R. 2012. *The Great American University: Its Rise to Preeminence, Its Indispensable National Role, Why It Must Be Protected*. New York: Public Affairs.

Cornelius, C. O. 1919. "Henry Hering's Sculptures for Field Museum of Natural History." *Field Museum of Natural History Annual Report* 5: 291-95.

Cullinane, S., H. Alkhshali, and M. Tawfeeq. 2015. "Tracking a Trail of Historical Obliteration: ISIS Trumpets Destruction of Nimrud." *CNN*, April 13.

Gillers, H. 2013. "Field Museum Cutting Costs, Losing Scientists." *Chicago Tribune*, July 31.

Gillers, H., and J. Grotto. 2013. "Dinosaur-Sized Debt: Field Museum Borrowed Heavily Before Recession; Now It's Paying the Price." *Chicago Tribune*, March 8.

Herzog, L. 2011. "Institutional Analysis of a Natural History Museum: Formation and Dissemination of Scientific Knowledge." Master's thesis, DePaul University, Chicago. [Sociological analysis of the Field Museum during the first decade of the twenty-first century. Herzog was an administrative assistant at the Field Museum from 1996 to 1998 and a fossil preparator from 1998 to 2012. She went back to graduate school for sociology while she was still a preparator at the museum. Today she is chief preparator at the North Carolina Museum of Natural Sciences.]

Hooper-Greenhill, E. 1992. *Museums and the Shaping of Knowledge*. London: Routledge Press.

Nugent, T. 2014. "Sue and the Bean Counter." *Nebraska Magazine* 110 (Spring): 34–39. [Article about Jim Croft, chief financial officer for the Field Museum from 1984 to present, with several quotes about the fiscal crisis of the Field Museum during the late 1990s and early 2000s.]

Shaheen, K. 2015. "Isis Fighters Destroying Ancient Artefacts at Mosul Museum." *Guardian*, February 26.

Shen, H. 2012. "Chicago's Field Museum Cuts Back on Science." *Nature Magazine*, December 20.

Webber, T. 2013. "Field Museum Debt Forces Reorganization." *Spokesman-Review*, July 6.

Figure Credits

Page 210: Compilation by Lance Grande based on four Field Museum images taken in 1918 by Charles Carpenter; Field Museum negative #CSGN40265, CSGN40270, CSGN40272, and CSGN40264.

Page 226: Photo by Lance Grande. Derringer is catalog number FMNH 279321, and *Cannabis* leaf is specimen number FMNH 1629791 from Afghanistan.

Page 227: Photo by Lance Grande.

Page 228: Photo by Karen Bean, 2009. Field Museum negative #GN91229_254CB.

Page 229: Photo by Lance Grande.

Page:230: Photo by Lance Grande, 2011.

Page 231: Photo by Alex Salmond's bodyguard.

CHAPTER TEN

Notes

1. The founding core of the Field Museum's gem collection came from Charles Tiffany (1812–1902) by way of the Higinbotham family. Tiffany was a groundbreaking business leader in New York City who in 1845 produced the first retail catalog in North America and by the late nineteenth century had created the continent's most prominent jewelry company. His company bought and sold some

of the world's finest gems, including some of the French crown jewels and pieces from the famous Hope collection. With the assistance of the famous gemologist George Kunz, Tiffany accumulated one of the world's finest private gem collections, which was exhibited in 1893 at the Chicago World's Fair. At the end of the six-month exhibition, the president of the World's Fair, Harlow Higinbotham, purchased the collection from Tiffany for $100,000, which was a very large sum of money back in 1893. Shortly afterward, he donated the collection to the new Chicago Natural History Museum (forerunner of the Field Museum of Natural History). The museum's first gem hall was the Higinbotham Hall of Gems, and Harlow Higinbotham went on to become the president of the museum from 1898 through 1908. A second major gem collection was later purchased from Tiffany's company by Harlow Higinbotham's daughter and presented to the museum in 1941.

Added Commentary

For further reading on gems, gemstones, and the Field Museum's Hall of Gems, see Grande and Augustyn (2009a). For a highly interactive app developed for the iPad, see Grande and Augustyn (2009b).

References

Grande, L., and A. Augustyn. 2009a. *Gems and Gemstones: Timeless Natural Beauty of the Mineral World*. Chicago: University of Chicago Press.

Grande, L., and A. Augustyn, 2009b. *Gems and Jewels*. iPad app. Touch Press and University of Chicago Press.

Figure Credits

Museum catalog numbers and additional information for specimens illustrated in this chapter are given in Grande and Augustyn (2009a).

Page 232: Photo by John Weinstein; Field Museum negative #A114779d_20B.
Page 245: Photo by Lance Grande, 2008.
Page 246: Photo by Lance Grande, 2008.
Page 247: Photos by John Weinstein and Lance Grande.
Page 248: Photo by John Weinstein and Lance Grande.
Page 249: Photo by John Weinstein and Lance Grande.
Page 250: Photo by John Weinstein and Lance Grande.
Page 251: Photos by John Weinstein and Lance Grande.
Page 252: Photo by John Weinstein and Lance Grande.
Page 253: Photo by John Weinstein and Lance Grande.

Page 254: Photo by John Weinstein.
Page 255: Photos by John Weinstein and Lance Grande.
Page 256: Photo by John Weinstein and Lance Grande.
Page 257: Photo by John Weinstein and Lance Grande.

CHAPTER ELEVEN

Notes

1. The collecting of human remains was not always for research or display. In the Middle Ages, there was a market for Egyptian mummies because they were thought to have medicinal properties. Mummy powder was sold by apothecaries for its supposed healing powers. Through the eighteenth century, the collecting of human remains and associated ethnographic and cultural materials was conducted largely by or for private collectors. By the nineteenth century, it had become a driving mission of museums around the Western world to preserve a record of human physical and cultural diversity.

2. The headline about Ota as the "Bushman" is from the *New York Times*, September 9, 1906.The tragic story of Ota Benga is told in a 2015 book by Pamela Newkirk called *Spectacle*. Ota was a young Congolese man exhibited in a cage at the New York zoo in 1906 and labeled as "The African Pygmy." He was a small man of 103 pounds standing only four feet eleven inches tall who was taken from Africa and presented to the director of the New York Zoological Gardens by failed missionary and businessman Samuel Phillips Verner. Verner had purchased Ota from African slavers for a pound of salt and a bolt of cloth. In the zoo exhibit, Ota was presented as an evolutionary "missing link," and people flocked to see the display. Ota was eventually rescued from the life of being treated as a zoo animal and embraced by African American communities in Brooklyn and Lynchburg, Virginia; but he continued to be haunted by a longing for his Congo home. He was never able to find a way back and finally committed suicide in 1916 at the age of thirty-two.

3. The concept of race typification in the early twentieth century continued to be an immensely popular concept in major museums around the world. In Chicago this eventually culminated in a major long-term exhibit called the *Hall of the Races of Mankind*. Planning for this exhibit first began in 1915, and it opened in 1933. The exhibit was extremely popular and attracted over 2 million visitors in the first year it was opened. It was not taken down until 1968, at which time the concept of racial typification and the context of the exhibit had become anathema to anthropologists. Most of the 116 sculptures from the exhibit representing a hundred different racial types by the famous sculptor Malvina Hoffman (student of Auguste Rodin) went into storage in the anthropology collection and eventually became extremely valuable art objects. In 2016 fifty of these sculptures were

put into a more humanistic exhibit called *Looking at Ourselves and Rethinking the Sculptures of Malvina Hoffman*. This exhibit revisits the concept of race, which has changed drastically over the last eighty years but is still very much with us. A 1911 quote from Franz Boas stands prominently on the wall of the new exhibit reading: "All races have contributed in the past to cultural progress. . . . They will be capable of advancing the interests of mankind if we are only willing to give them a fair opportunity."

4. Helen A. Robbins, PhD in anthropology from the University of Arizona, 2001.

References

Anon. 2000. From the Photo Archives. *In the Field* 71 (July-August): 12. [Short story about the murder of assistant curator William Jones on Luzon Island].

Boas, F. 1974. *A Franz Boas Reader: The Shaping of American Anthropology, 1883-1911*, ed. G. W. Stocking Jr. Chicago: University of Chicago Press.

Forde, C., J. Hubert, and P. Turnbull, eds. 2002. *The Dead and Their Possessions: Repatriation in Principle, Policy and Practice*. New York: Routledge Taylor & Francis Group.

Gossett, T. F. 1963. *Race: The History of an Idea in America*. New York: Oxford University Press. [2nd ed., 1997.]

Landau, P. M., and D. G. Gentry Steel. 1996. "Why Anthropologists Study Human Remains." *American Indian Quarterly* 20: 209-28.

McManamon, F. P. 1992. "Managing Repatriation: Implementing the Native American Graves Protection and Repatriation Act." *CRM Bulletin* 15: 9-12.

Nash, S., and G. Feinman, eds. 2003. "Curators, Collections, and Contexts: Anthropology at the Field Museum, 1893-2002." *Fieldiana, Anthropology*, n.s., no. 36, Publication No. 1525. [Recommended reading for history of the Field Museum's anthropology department.]

Newkirk, P. 2015. *Spectacle: The Astonishing Life of Ota Benga*. New York: Amistad/HarperCollins.

Owsley, D., and K. Bruwelheide. 2009. *Written in Bone: Bone Biographer's Casebook*. Minneapolis: LeanTo Press.

Page, K. 2011. "The Significance of Human Remains in Museum Collections: Implications for Collection Management." Master's thesis, State University of New York, Buffalo.

Robbins, H. 2014. "In Consideration of Restitution: Understanding and Transcending the Limits of Repatriation under the Native American Graves Protection and Repatriation Act (NAGPRA)." In *Museums and Restitution: New Practices, New Approaches*, ed. L. Tythacott and K. Arvanitis, 105-18. Farnham, Surrey, UK: Ashgate Press.

Rose, J. C., T. J. Green, and V. Green. 1996. "NAGPRA IS FOREVER: Osteology and the Repatriation of Skeletons." *Annual Review of Anthropology* 25: 81-103.

Figure Credits

Page 258: Photographer unknown; Smithsonian Institution Archives negative #MNH8304.
Page 277: Photo by Stephen Simms; Field Museum negative #CSA33663_A.
Page 278: Photo by Mark Widhalm, 1998; Field Museum negative #GN88941_10c.
Page 279: Photo by Lance Grande, 2014.
Page 280: Photo by John Weinstein; Field Museum negative #GN91515_007d.
Page 281: Photos by Helen Robbins.
Page 282: Photo by Karen Bean; Field Museum negative #GN91537_122d.
Page 283. Images by J. P. Brown, 2011.

CHAPTER TWELVE

Notes

1. The man-eaters' diorama is one of the things that first drew Tom Gnoske to the Field Museum, at the young age of four. Fittingly, it was he who in 1997 figured out how to find the lost man-eaters' cave that John Henry Patterson had discovered nearly a century before. Once Gnoske and his colleagues relocated Patterson's cave and reported that there were no human bones inside of it, a controversy started over whether or not the cave could have been a lion's den. But it was in an area that had been repeatedly flooded during El Niño years over the last century. Any bones from the man-eaters in 1899 would have long since been washed away. Some scientists argued that lions do not live in caves today. But then in 1998, Gnoske and Kerbis-Peterhans found a family of cave-dwelling lions living in neighboring Uganda, the ultimate end of the Uganda Railroad line. A final challenge to the cave's legitimacy as the man-eaters' den came from Chap Kusimba, who suggested the possibility that what Patterson had seen may have been a burial shrine rather than a lion's den. But because the original bones were no longer there, there was no way to test this hypothesis. The controversy surrounding Colonel Patterson's cave has still not been completely resolved, and in fact may never be.

2. Bruce D. Patterson, PhD in biology from New Mexico University, 1981. Bruce authored a popular book entitled *The Lions of Tsavo* (McGraw-Hill, 2004).

References

Aderet, O. 2014. "Ashes of WWI Jewish Legion Chief to Be Interred in Israel." *Jewish World News*, October 13. http://www.haaretz.com/jewish/news/.premium-1.620433. [News report about John Patterson's ashes being transferred to Israel for internment.]

Anthony, G., J. Estes, M. Ricca, A. Miles, and E. Forsman. 2008. "Bald Eagles and Sea Otters in the Aleutian Archipelago: Indirect Effects of Trophic Cascades." *Ecology* 89: 2725-35.

Conrad, E. C. 1982. "Are There Fossils in the 'Wrong Place' for Evolution?" *Creation Evolution Journal* 3: 14-22.

Gnoske, T., and J. Kerbis-Peterhans. 2000. "Cave Lions: The Truth Behind Biblical Myths." *In the Field*, 71: 2-6.

Guilford, G. 2013. "Why Does a Rhino Horn Cost $300,000? Because Vietnam Thinks It Cures Cancer and Hangovers." *The Atlantic*, May 15.

Kerbis-Peterhans, J. C., C. M. Kusimba, T. P. Gnoske, S. Andanje, and B. D. Patterson. 1998. "Man-Eaters of Tsavo Rediscovered after 100 Years, an Infamous 'Lion's Den,' Rekindles Some Old Questions." *Natural History* 107: 12-14.

Leakey, M. G., C. S. Feibel, I. MacDougall, and A. Walker. 1995. "New Four-Million-Year-Old Hominid Species from Kanapoi and Allia Bay, Kenya." *Nature* 376, no. 6541: 565-71.

Leakey, M., and A. Walker. 1997. "Early Hominid Fossils from Africa." *Scientific American*, June.

Ngene, S., E. Bitok, J. Mukeka, F. Gakuya, P. Omondi, K. Kimitie, Y. Watol, L. Kariuki, and O. Okita. 2011. "Census and Ear-Notching of Black Rhinos (*Diceros bicornis michaeli*) in Tsavo East National Park, Kenya." *Pachyderm*, no. 49: 61-69.

Olsen, E. C. 1985. "Bryan Patterson, 1909-1979." *Biographical Memoirs*, National Academy of Sciences: 435-50.

Patterson, Bruce. 2004a. *The Lions of Tsavo: Exploring the Legacy of Africa's Notorious Man-Eaters*. New York: McGraw-Hill.

———. 2004b. "Maneless and Misunderstood: The Lions of Tsavo." *Earthwatch Institute* 23: 12-15.

Patterson, Bruce, S. Kasiki, E. Selempo, and R. Kays. 2004. "Livestock Predation by Lions (*Panthera leo*) and Other Carnivores on Ranches Neighboring Tsavo National Parks, Kenya." *Biological Conservation* 119: 507-16.

Patterson, Bryan, and N. W. Howells. 1967. "Hominid Humeral Fragments from the Early Pleistocene of North-Western Kenya." *Science* 156: 64-66.

Patterson, J. 1907. *The Man-Eaters of Tsavo and Other East African Adventures*. London: MacMillan.

Sheldrick Wildlife Trust. 2014. *Tsavo Ecosystem Elephant Count*. https://www.sheldrickwildlifetrust.org/updates/updates.asp?ID=616.

Von Buol, P. 1998. "Into the Man-Eater's Den." *Safari*, August/September, 9-17.

Wagner, J. "A Wolf's Role in the Ecosystem—the Trophic Cascade," *Mission: Wolf*. http://www.missionwolf.org/page/trophic-cascade/.

Figure Credits

Page 284: Photo by John Weinstein, Field Museum negative #Z94352c.

Page 299: (*Top*) Photographer unknown; Field Museum negative #Z93658; (*bottom*) Photographer unknown; Field Museum negative #Z93657.

Page 300: Photo by Lance Grande.

Page 301: (*Top*) Photographer unknown; Field Museum negative #GEO80350; (*bottom*) From the Archives of the Museum of Comparative Zoology, Ernst Mayr Library, Harvard University.

Page 302: (*Top*) Photo by John Weinstein, Field Museum negative #Z983932c; (*bottom*) Photo by Bruce Patterson.

Page 303: Photo by J. P. Brown and Bruce Patterson.

Page 304: Photo by Lance Grande.

Page 305: Photo by Lance Grande.

CHAPTER THIRTEEN

Notes

1. Lawrence R. Heaney, PhD in systematics and ecology from the University of Kansas, 1979.

2. Rüdiger Bieler, PhD in zoology, University of Hamburg, 1985.

3. Michael J. Novacek, PhD in paleontology, University of California, Berkeley, 1978.

4. Eleanor J. Sterling, PhD in forestry, environmental studies, and anthropology, Yale University, 1993. Sterling is a world authority on nocturnal lemurs in Madagascar and a major authority on the biodiversity of tropical regions all over the world. Her training in both environmental studies and anthropology enabled her to combine biodiversity research with consideration of local peoples in the areas of concern to sustain conservation initiatives long after the CBC leaves. In 2014 Eleanor stepped down from the directorship of the CBC and took on a new role as its Chief Conservation Scientist, to advance key CBC research and global outreach initiatives. Ana Luz Porzecanski (PhD in ecology and evolutionary biology, Columbia University, 2003) took over the directorship. This brought things full circle in some ways, because Ana was a PhD student of curator of ornithology, Joel Cracraft, who was a key member of the team that originally developed the CBC and was its interim director in 1994.

5. Debra K. Moskovits, PhD in biology, University of Chicago, 1985.

6. Information on staffing levels taken from Field Museum website.

7. See fieldmuseum.org/search/google/field%20guides?query=field%20guides.

8. Robin Foster, PhD in Botany, Duke University, 1974.

9. Alaka Wali, PhD in anthropology, Columbia University, 1984.

References

Catibog-Sinha, C., and L. R. Heaney. 2006. *Philippine Biodiversity: Principles and Practice*. Diliman, Philippines: Haribon Foundation.

Foster, R. 1977. "*Tachigalia versicolor* Is a Suicidal Neotropical Tree." *Nature* 268: 624–26.

Heaney, L. R., ed. 2011. "Discovering Diversity: Studies of the Mammals of Luzon Island, Philippines." *Fieldiana, Life and Earth Sciences*, no. 2.

Heaney, L. R., and D. S. Balete. 2012. "Discovering Diversity: Newly Discovered Mammals Highlight Areas of Unique Biodiversity on Luzon." *Wildernews* 1: 4–9.

Heaney, L. R., and S. M. Goodman. 2009. "Mammal Radiations." In *Encyclopedia of Islands*, ed. R. Gillespie and D. Clague, 588–91. Berkeley: University of California Press.

Heaney, L. R., and J. C. Regalado Jr. 1998. *Vanishing Treasures of the Philippine Rain Forest*. Chicago: Field Museum.

Jansa, S., K. Barker, and L. R. Heaney. 2006. "The Pattern and Timing of Diversification of Philippine Endemic Rodents: Evidence from Mitochondrial and Nuclear Gene Sequences." *Systematic Biology* 55: 73–88.

Kolbert, E. 2014. *The Sixth Extinction*. New York: Henry Holt.

MacNamara, P., J. M. Bates, and J. H. Boone. 2008. *Architecture by Birds and Insects: A Natural Art*. Chicago: University of Chicago Press.

Mikkelsen, P. M., and R. Bieler. 2007. Seashells of Southern Florida—Living Marine Mollusks of the Florida Keys and Adjacent Regions: Bivalves. Princeton, NJ: Princeton University Press.

Moskovits, D. K., C. Fialkowski, G. M. Mueller, and T. A. Sullivan. 2002. "Chicago Wilderness: A New Force in Urban Conservation." *Annals of Missouri Botanical Garden* 89: 153–63.

Novacek, M. 2007. *Terra: Our 100-Year-Old Ecosystem—and the Threats That Now Put It at Risk*. New York: Farrar, Straus & Giroux.

Figure Credits

Page 306: Photo by Rafe Brown.
Page 322: Photo by Danilo Balete, 2007.
Page 323: Image by Kasey Mennie and Lance Grande.
Page 324: Photo by Danilo Balete.
Page 325: Photos by Danilo Balete, 2011.
Page 326: Photos by Larry Heaney.
Page 327: Photos by Danilo Balete.
Page 328: Photos by Danilo Balete.
Page 329: Photo by Petra Sierwald.
Page 330: Photo by Petra Sierwald, 2014.

Page 331: Photos by Petra Sierwald.
Page 332: Photo courtesy of Eleanor Sterling. Photographer: Thach Mai Hoang.
Page 333: Photo by Alvaro del Campo.
Page 334: Photo by Alvaro del Campo.
Page 335: Photo courtesy of Robin Foster, 2010.
Page 336: Photo by Diana Alvira, 2009.
Page 337: Image by Jon Markel, somewhat modified by Lance Grande.

CHAPTER FOURTEEN

Notes

1. The two main types of research in museums are basic research and applied research, although there is some overlap. Applied research answers specific questions that have direct and immediate applications, ranging from medical research (such as the pathogens project mentioned in chapter 7) to environmental conservation (chapters 12 and 13). Basic research seeks simply to increase our general scientific knowledge base. Curatorial research often prioritizes specimen-based basic research driven purely by curiosity and the desire to expand our knowledge of the natural and cultural sciences. The first stated mission of the Field Museum website in 2014 reads: "To inspire curiosity and understanding about life on Earth" (i.e., basic research). Basic research eventually leads to applications answering some of the biggest questions in science, often bringing huge benefits to society that were never originally imagined. Basic research historically led to a range of invaluable discoveries, from laptop computers and iPhones to DNA and lasers. Many advances in medicines to fight heart disease, Alzheimer's disease, and even cancer have resulted in part from basic research on plant and animal diversity. The eventual applications of basic research studies are often not fully appreciated until decades later. If the only type of research done was applied research, then humans would still be stuck on developing ways to make better spears. And basic research in natural and cultural history results in a better understanding of our place in the network of Earth's biodiversity and cultural complexity. That understanding is a measure of our worth as a species and fundamental to strategizing our future on this planet.

2. The current culture of academia (including both research curators and university professors) is the result of a historical process in which my own career developed. This makes my perspective the subjective one of an insider. The years that I spent in graduate school in New York and my first two decades in Chicago occurred during a golden age for scientific research institutions. The 1980s brought a surge of funding from federal agencies for research, and there was

an enhanced effort to train research scientists. The pool of unemployed and underemployed PhD graduates from the 1970s was drying up. According to history professors Hugh Graham and Nancy Diamond (1997), the number of employed academic scientists increased by 41 percent between 1978 and 1990. The National Science Foundation, established in the 1950s, and other federal agencies were supporting scientific research institutions (mostly universities and research museums) at unprecedented levels, giving major grants to individual professors and curators of those institutions who were doing basic research. This allowed the institutions to leverage their investment in research scientists into broad, cutting-edge scientific programs. These programs generated new discoveries and scientific advances that enhanced the reputations of research institutions and the country as a whole. On a more personal level, they also served to build my own scientific career.

The enhanced availability of federal funds allowed universities and museums to expand their research programs in fiscally prudent ways. Federal grants come with overhead (also called indirect costs) to the scientist's institution. This amount is in addition to the added costs of the project itself. Indirect cost rates recovered by institutions can be significant. For example, Harvard's indirect cost rate for federally supported on-campus research (posted on its website) was 69 percent in 2014. This meant that for a Harvard researcher to write a grant proposal to the National Science Foundation or the National Institutes of Health to fund a $1 million research project, he or she could also include an additional $690,000 for the university to cover overhead costs (expenses related to the overall infrastructure of the building, administrative costs, and even part or all of the researcher's salary). The added funding for research universities and research museums resulted in their becoming the main source of this country's scientific and technological preeminence by the late twentieth century. Federal grant support for research resulted in major discoveries, particularly in areas of medical research and physics (e.g., discoveries ranging from DNA structure to the development of MRI imaging), as well as in systematic biology (deciphering the tree of life on Earth).

Although major grant money for research was now available, the process of winning it became extremely competitive. In many programs (e.g., systematic biology) only about 10 to 14 percent of the proposals were funded at best, and nearly all of the proposals submitted were very good to excellent ones. Each submitted proposal represented a lengthy and time-consuming effort. Some of my own proposals (including appendices) ranged from 60 to 150 pages each and took a month or more to prepare. It constituted a risk for the scientist and institution to invest the time and resources to prepare such proposals. As a result, research institutions had to focus more than ever on having the best scientists available in order to get the most out of their research investment: not only in terms of ramping up scientific discoveries to enhance an institution's reputation, but also in terms of getting grants to help defray the cost of housing major research programs.

As academic institutions increased their expectations of research scientists, requirements to get the top research jobs became increasingly competitive for applicants. By the late twentieth century, a PhD had become the minimum educational requirement for a professorship or curatorship in a major university or museum. Applicants for these positions were increasingly evaluated on readily measurable units of productivity (grants and publications). As PhD students in the late 1970s and early 1980s, it was pounded into our heads by our graduate advisors that the main requirements to be a successful curator or professor in a major research institution were to publish (or perish), to get federal grants (including at least one major federal grant in the first few years in order to get tenure), and in the case of museums also to develop a productive field program. The major research institutions incorporated those criteria into their policies for hiring, tenure decisions, and promotion. Other factors such as service to the institution, public outreach, or teaching were given much less weight, even in major universities (e.g., see discussion of Harvard in chapter 2, note 1, on page 368). In order to satisfy the hurdles to achieve promotion, researchers had to focus their energies on successfully communicating with other scientists within the professional community. The unintended consequence of this was that it left little time for researchers to write to the general public about the importance of the work they were doing. I believe that a future challenge for senior curators in major museums will be to better balance technical research with public outreach. The challenge for museum executive management will be to help them do that.

3. Evolutionary theory should not be in conflict with personal spirituality. It is just different. Apples and oranges. Many scientists have some sort of religious faith or spirituality. One of my favorite quotes from the philosopher Karl Popper, appealing to Einstein's way of expressing himself in theological terms, is "If God had wanted to put everything into the universe from the beginning, He would have created a universe without change, without organisms and evolution, and without man and man's experience of change. But he seems to have thought that a live universe with events unexpected even by Himself would be more interesting than a dead one" (as quoted from Hartshorne 1984).

4. One of the fastest-growing efforts to disseminate curatorial research and fieldwork to the general public today relies on the Internet and social media. An early effort by the Field Museum to deliver curatorial content more widely through the Internet was a website called Expeditions at the Field (http://expeditions.fieldmuseum.org/) funded by the Negaunee Foundation. It was a modest pioneering attempt that although successful in drawing thousands of viewers, it drew the interest of people who were already online looking for such things. Then, a few years later while heading the C&R division, I hired a videog-

rapher and media producer, Federico Pardo, to create a second series called The Field Revealed. Unlike the interactive Expeditions site that focused primarily on fieldwork, The Field Revealed site was a series of short videos that spotlighted collection-based research projects of the museum (https://www.fieldmuseum.org/topic/field-revealed). This added to our online viewership, as did individual curators developing independent blog sites (e.g., Michael Dillon's www.sacha.org). The Field Museum's most successful social media investment to date has been hiring the talented American science communicator and YouTube educator Emily Graslie as our full-time chief curiosity correspondent. She stars in her own educational YouTube channel called The Brain Scoop, and she has done many segments on curatorial research programs in the museum. By the end of 2014, after just two years, the channel had 10.2 million video views and 270,000 subscribers. Emily and her producer Tom McNamara did one feature with me on gems and gemstones (https://www.youtube.com/watch?v=zcHy0oo7QU0). They also came out with me, my field crew, and my "Stones and Bones" students to Wyoming in 2014 to produce some episodes about my field site and research (e.g., http://thebrainscoop.tumblr.com/post/93341370471/the-brain-scoop-fossil-fish-pt-ii-a-history-52). Keeping up with the quickly evolving Internet technology is another one of the big challenges for museums and curators in the twenty-first century. Today we need to reach people not only through their laptops; we need to connect to handheld devices such as smartphones, smart watches, and other new devices yet to come into public use. We also need to keep up with the ever-changing types of social media platforms.

5. There are curators who have been extremely active and successful in addressing the wider public (examples referenced below), including Margaret Mead and Niles Eldredge (curators at the American Museum of Natural History), Louis Leakey (curator at the Coryndon Memorial Museum in Nairobi), Stephen Jay Gould and E. O. Wilson (curators at the Museum of Comparative Zoology), and Richard Fortey (a curator at the Natural History Museum in London). Other scientists who are particularly successful in addressing the general public, include Neil deGrasse Tyson (Rose Center for Earth and Space at the American Museum of Natural History) and my colleague Neil Shubin (University of Chicago). All of these outstanding educators have done much to heighten society's appreciation of science and to encourage public support of basic research. But we need more such communicators today.

References

Adkins, R. 2012. "America Desperately Needs More STEM Students: Here's How to Get Them." *Forbes* online, July 9.
Alliance of Natural History Museums of Canada. 2011. *A National Collections*

Strategy for Canada's Natural History Museums. http://www.natural historymuseums.ca/documents/ANHMC_English_StrategyPaper.pdf.

Atkinson, R., and W. Blanpied. 2008. "Research Universities: Core of the U.S. Science and Technology System." *Technology and Society* 30: 30–48.

Chappell, B. 2013. "U.S. Students Slide in Global Ranking on Math, Reading and Science." *NPR: The Two-Way. Breaking News from NPR.* December 3. http://www.npr.org/sections/thetwo-way/2013/12/03/248329823/u-s-high-school-students-slide-in-math-reading-science.

Cole, J. R. 2012. *The Great American University: Its Rise to Preeminence, Its Indispensable National Role, Why It Must Be Protected.* New York: Public Affairs.

Duncan, D. 2007. "216 Million Americans Are Scientifically Illiterate (Part I)." *MIT Technology Review*, February 21.

Eldredge, N. 2000. *The Triumph of Evolution and the Failure of Creationism.* New York: W. H. Freeman and Co.

———. 2005. *Darwin: Discovering the Tree of Life.* New York: W. W. Norton.

Fabricant, D. S., and N. Farnsworth. 2001. "The Value of Plants Used in Traditional Medicine for Drug Discovery." *Environmental Health Perspectives* 109: 69–75.

Fandos, N., and N. Pisner. 2013. "Joining the Ranks. Demystifying Harvard's Tenure System." *The Harvard Crimson*, April 11.

Finn, R. 2001. "Snake Venom Protein Paralyzes Cancer Cells." *Journal of the National Cancer Institute* 93, no. 4: 261–62.

Fortey, R. 1999. *Life: A Natural History of the First Four Billion Years of Life on Earth.* New York: Vintage Books.

———. 2005. *Earth: An Intimate History.* New York: Vintage Books.

———. 2008. *Dry Storeroom No. 1: The Secret Life of the Natural History Museum.* London: Harper.

Gould, S. J. 1991. *Wonderful Life: The Burgess Shale and the Nature of History.* New York: W.W. Norton. [Finalist for the Pulitzer Prize in 1991 and a national best seller, this book's thesis was that chance was one of the decisive factors in the evolution of life on Earth.]

Graham, H. D., and N. Diamond. 1997. *The Rise of American Research Universities: Elites and Challengers in the Post-War Era.* Baltimore: Johns Hopkins University Press.

Hartshorne, C. 1984. *Omnipotence and Other Theological Mistakes.* Albany: State University of New York Press.

Haynie, D. 2013. "U.S. Sees Record Numbers of International College Students." *U.S. News and World Report*, online, November 11.

Hölldobler, B., and E. O. Wilson. 1991. *The Ants.* Cambridge, MA: Belknap/Harvard University Press. [Winner of the 1991 Pulitzer Prize, this book is a sweeping survey of the ethnology, ecology, and evolutionary biology of ants.]

Jensen, P., J.-P. Rouquier, P. Kreimer, and Y. Croissant. 2008. "Scientists Who

Engage with Society Perform Better Academically." *Science and Public Policy* 35: 527–41.

Kehoe, A. 1998. "Why Target Evolution? The Problem of Authority." In *Scientists Confront Creationism: Intelligent Design and Beyond*, ed. A. Petto and L. Godfrey, 381–404. New York: W. W. Norton.

Leakey, L., and V. Goodall. 1969. *Unveiling Man's Origins*. Cambridge, MA: Schenkman Publishing.

Mahidol, C., S. Ruchirawat, H. Prawat, S. Pisutjaroenpong, S. Engprasert, P. Chumsri, T. Tengchaisri, S. Sirisinha, and P. Picha. 1998. "Biodiversity and Natural Product Drug Discovery." *Pure and Applied Chemistry* 70: 2065–72.

Mead, M. 1972. *Blackberry Winter: My Earlier Years*. New York: William Morrow.

Morris, H. 1974. *The Troubled Waters of Evolution*. Master Books.

Newport, F. 2014. "In U.S., 42% Believe Creationist View of Human Origins." Gallup online, June 2.

Patlak, M. 2003. "From Viper's Venom to Drug Design: Treating Hypertension." *FASEB Breakthroughs in Bioscience*. http://www.fasebj.org/cgi/reprint/18/3/421e.

Petto, A., and L. Godfrey. 1998. *Scientists Confront Intelligent Design and Creationism*. New York: W. W. Norton.

"PISA 2012 Results," OECD, http://www.oecd.org/pisa/keyfindings/pisa-2012-results.htm.

Powers, K., L. Prether, J. Cook, J. Woolley, H. Bart, A. Monfils, and P. Sierwald. 2014. "Revolutionizing the Use of Natural History Collections in Education." *Science Education Review* 13: 24–33.

Ramsey, G. F. 1938. *Educational Work in Museums in the United States: Development, Methods, and Trends*. New York: H. W. Wilson.

Reese, M. 2006. *Science Communication: Survey of Factors Affecting Science Communication by Scientists and Engineers*. London: The Royal Society. https://royalsociety.org/~/media/Royal_Society_Content/policy/publications/2006/1111111395.pdf.

Schudel, M. 2006. "Obituary: Henry M. Morris, Father of 'Creation Science.'" *Seattle Times*, March 5, 2006.

Shubin, N. 2008. *Your Inner Fish: A Journey into the 3.5-Billion-Year-Old History of the Human Body*. New York: Pantheon Books.

Tewksbury, J., J. Anderson, J. Bakker, T. Billo, P. Dunwiddie, M. Groom, S. Hampton et al. 2014. "Natural History's Place in Science and Society." *BioScience* 20: 1–11.

Tyson, N. d. 2014. *Cosmos: A Spacetime Odyssey*. [A documentary TV series that is also on DVD and Blu-ray.]

Wilson, E. O. 1979. *On Human Nature*. Cambridge, MA: Harvard University Press. [Winner of the 1979 Pulitzer Prize; a presentation of how different characteristics of human society can be explained as a result of evolution.]

Figure Credits

Page 338: Photo by John Weinstein for L. Grande. Time-lapse photo of about five minutes.

Page 349: Photo by John Weinstein; Field Museum negative #GN91729_18d.

Page 350: (*Top*) Photo by John Weinstein, 2003; Field Museum negative #GN90605_59D; (*bottom left*) Photo by Mark Widhalm, 2004; Field Museum negative #GN90679_22D; (*bottom right*) Photo by John Weinstein, 2003; Field Museum negative #GN90605_04D.

Page 351: (*Top*) Photo by L. Grande, 2015; (*bottom left*) Photo by Lance Grande, 2015; (*bottom right*) Photo by John Weinstein, 2007; Field Museum negative #GN91082_015d.

Page 352: Photo by Lance Grande.

INDEX

acid-transfer technique for preparing fossils, 4, 5, 15
Africa, 162, 177–79, 185, 187, 195, 284–305
"Agusan Gold Image," 241, 242, 256
Alabama Deep Sea Fishing Rodeo, 95–100, 104–7
Alvin deep-sea submersible, 167, 168, 188
Angielczyk, Ken (curator of vertebrate paleontology, Field Museum), 166, 167, 187, 382
Antarctica, 138, 163, 164, 186
Applegate, Shelly (curator of paleontology, National Autonomous University of Mexico), 65–78, 81, 374
applied research (definition), 400
Aranguthy family, 67, 71–73, 79–81
Augustyn, Allison, 238, 244, 245
Aztec "Sun-god Opal," 240, 255

Bakker, Robert (curator of paleontology, Houston Museum of Natural Science), 115–17, 119, 135, 378
Bardack, David, 23, 66–74, 79

Bates, John (curator of birds, Field Museum), 162, 163, 185
Bemis, Willy (curator of zoology, Cornell Museum of Vertebrates), 84–109, 375, 376
Bergelson, Joy, 391
Bieler, Rüdiger (curator of invertebrates, Field Museum), 313, 314, 398
biogeography, 358
bird collection, 162, 163, 185, 382
Black Hills Institute, 113–31, 140–46, 378, 379
Boas, Franz (curator of anthropology, Field Museum; Smithsonian's National Museum of Natural History; and American Museum of Natural History), 258, 260–66, 268, 395
Bolt, John (curator of vertebrate paleontology, Field Museum), 24, 27, 28, 30, 33, 135, 369
Boudreaux, Gail, 354
Boudreaux, Terry, 354

Boyd, Willard (Sandy) (president of Field Museum, 1981–96), 123–25, 317, 379
Brake, Jamie, 280
Brennan, Joseph, 272
Brochu, Chris, 137, 379
Brown, Barbara, 354
Brown, J. P., 275, 282
Brown, Roger, 354
Brundin, Lars, 20, 363, 364
Buchheim, Paul, 46

cabinet of curiosities, xi
Canning, John, 354
Canning, Rita, 354
Carhart, Drew, 63, 352
Center for Biodiversity Conservation (CBC), 314–16, 332, 398
Chaifetz, Jill, 354
Chaifetz, Richard, 354
Chalifa, Yael, 90, 91
challenges for natural history museums, 213, 219, 220, 276, 310, 320, 340–47, 381–84, 402, 403
citizen-science, 42, 43, 97, 162, 217, 294, 347, 390
cladistics, 7–13, 18–20, 357, 358, 363, 364
cladograms, 9, 18, 19
Clinton, Bill, 138, 152
cloud rats, 311, 322, 326
Coats, Michael, 391
Colburn, Richard, 353
Colburn, Robin, 353
collecting animals for museum collections, 387
collections-based research (also called specimen-based research), x, 29, 389, 400
collection size, 355
commercial fossil operations 36–46. *See also* Black Hills Institute
Committee on Evolutionary Biology (CEB), 26, 216
consortium of natural history museums, 341
coral reef restoration, 313, 314, 329–31

Cracraft, Joel (curator of birds, American Museum of Natural History), 315, 398
Crane, Peter (curator of paleobotany, Field Museum; Dean of Forestry and Environmental Studies, Yale University), 28, 30, 31, 132, 133, 147, 316, 317, 369, 370
creationism and young Earth creationists, xiii, 12, 293, 343–45, 362
CT scanning, 120, 134, 139, 154, 155, 275, 276, 282, 283
curator, adjunct (definition), 368, 369
curator, assistant (definition), 368
curator, associate (definition), 368
curator, emeritus (definition), 369
curator, full (definition), 368
curatorial ranks, 368
curator of exhibits positions in natural history museums, xiv
curators, as percentage of total museum staff members, 381

dermestid beetles (as skeletonizing organisms), 99, 106, 107, 163, 185, 374–76
dicynodonts, 166, 167
Dillon, Michael (curator of botany, Field Museum), 161, 162, 183, 184, 382
Dingerkus, Guido (curator of ichthyology, Muséum national d'histoire naturelle), 357
Dinosaur 13 (movie), 112–14
DNA, 189, 351
Dorsey, George (curator of anthropology, Field Museum), 266, 267, 277

Egypt, 217–19, 230, 259, 275, 282, 283, 394
Eimer, Nathan, 125–27, 379
Eklund, Mike, 63, 352
Eldredge, Niles (curator of fossil invertebrates, American Museum of Natural History), 130, 403
Encyclopedia of Life (EOL), 216–20, 230, 390

Environment, Culture, and Conservation (ECCo), 316-20, 333-37
environmental conservation, 295-98, 302, 304-37, 341, 400
evolutionary pattern vs. evolutionary process, 358, 359
evolutionary research, 8-10, 18, 169, 344, 345, 358, 359, 361-63, 402
Evolving Planet exhibit, 47
exhibitions, xiv, xv, 46, 135-37, 148-52, 232-57, 347
exigency committee, 223, 224
Expeditions at the Field website, 402
extinction, xiii, 44, 167, 298, 307, 316, 317, 320, 360, 387

Father of American Anthropology. *See* Boas, Franz
federal agencies (interacting with, as head of collections and research), 214-16
federal funding for research, 66, 86, 222, 342, 400-402
Feinman, Gary (curator of anthropology, Field Museum), 172, 173, 191, 384
Field Revealed website, 403
fiscal challenges for Field Museum, 220-24, 340
flagship species, 298, 304, 305
Florida National Marine Sanctuary, 313, 314, 329-31
Flynn, John (curator of vertebrate paleontology, Field Museum, and American Museum of Natural History), 28, 30, 33, 132, 370
Foote, Michael, 390
Fossil Butte National Monument, consulting on, 46
fossil meteorites, 165, 166, 187
fossil record (incompleteness of), 360, 361
Foster, Robin (adjunct curator of botany, Field Museum), 319, 335, 398
Fraley, Phil, 136, 137
frozen tissue collection, 340, 351
fugu (pufferfish), 93, 94

Gantz, Bill, 354
Gantz, Linda, 354
gem hall, Field Museum, 232-57, 392, 393
Gieser, Ellen, 63, 352
Gingrich, Newt, 31
Gnoske, Tom, 290, 291, 293, 295, 396
Gould, Stephen Jay (curator of invertebrate paleontology, Museum of Comparative Zoology, Harvard), 167, 403
Graham School of the University of Chicago, 47, 62, 348, 352
Grainger, David, 253, 254
Grainger, Juli, 253, 254
Grainger Foundation, 253
Grainger Hall of Gems, 232-57
Graslie, Emily (YouTube personality), 352, 403
Grauer, Anne, 279

Haas, Jonathan (curator of anthropology, Field Museum), 268, 269, 278
Hackett, Shannon (curator of birds, Field Museum), 163, 169, 189
Hans. *See* Radke, Eric
Heaney, Larry (curator of mammals, Field Museum), 308-13, 322-28
Heck, Philipp (curator of meteoritics, Field Museum), 165, 166, 187, 382
Hendrickson, Susan, 115-17, 130, 135, 138, 141, 143, 149
Hennig, Willi, 364
herbarium, 159, 160, 181
Hilton, Eric, 108
Holmes, William Henry (curator of anthropology, Field Museum), 264, 266
Holstein, Jim, 63, 229, 352
human remains, 258-83, 394
human zoos, 262, 394
hypothesis (in science), 359

indirect costs (expenses of doing business), for museums, 213, 221, 340
indirect costs in federal grants, 401

Index   409

Inger, Robert (curator of herpetology, Field Museum), 201
Inuit people, 261, 262, 270–73, 280, 281

Jackson, Rick, 35, 36, 39, 40
Japan, 92–94, 103
jewelry designers (who worked with me on the Grainger Hall of Gems), 243, 244
Johnson, Kirk (curator of paleontology, Denver Museum of Natural History; director and CEO of Smithsonian's National Museum of National History), 382
Jones, William (curator of anthropology, Field Museum), 267, 277

Kenya, 177–79, 195, 297–300, 302, 304, 305
Kerbis-Peterhans, Julian (adjunct curator of mammals, Field Museum), 291, 293, 295, 396
Kusimba, Chapurukha (Chap) (curator of anthropology; Field Museum), 177–79, 195, 291, 293, 300, 384, 396

Lampert, Lester, 243, 244, 247, 251
Lariviere, Richard (president of Field Museum, 2012–present), 224, 353
Larson, Bill, 238
Larson, Peter, 113–30, 123, 129–31, 135, 140, 142, 143
laws (in science), 359
Lewis Ranch, Wyoming, 40, 49–51, 62, 63, 352
Library of Alexandria, 218, 219
lichens, 160, 161, 182
Lidgard, Scott (curator of invertebrate paleontology, Field Museum), 28, 30, 33, 370
Linnaeus, Carolus, xii, xiii, 355, 356
Lücking, Robert (adjunct curator of botany, Field Museum), 160, 161, 182
Lumbsch, Thorsten (curator of botany, Field Museum), 160, 161, 180, 182

MacArthur Foundation, 353
Makovicky, Peter (curator of vertebrate paleontology, Field Museum), 138, 139, 153, 154, 379
man-eaters of Tsavo (lions), 284–300, 303, 396; cave of, 290, 291, 300, 396
Martin, Robert (curator of anthropology, Field Museum), 176, 177, 194, 275, 384
Marx, Hymen (curator of herpetology, Field Museum), 201
Masek, Bob, 148
Mayr, Ernst (curator of ornithology, American Museum of Natural History and Museum of Comparative Zoology, Harvard), 12, 361, 363
McCarter, John (president of Field Museum, 1996–2012), 131–33, 147, 152, 233, 271, 350, 379
Mead, Margaret (curator of anthropology, American Museum of Natural History), 264, 269, 382, 403
meteorites, 163–66, 186, 187, 221, 222, 228, 229
Mexico, 64–83
mining for gems, 237–39, 245, 246
mission statement, Field Museum, 389, 390, 400
Mitchell, John, 63, 352
Moreau, Corrie (curator of entomology, Field Museum), 170–72, 190, 382, 384
Morrill, Bryan, 63, 352
Moscow, 86–89, 102
Moskovits, Debra, 317, 333, 398
Mueller, Greg (curator of mushrooms and fungi, Field Museum; Vice President of Science, Chicago Botanic Garden), 158, 159, 180, 382
mummies, 259, 265, 275, 276, 282, 283, 394
mushrooms, 158, 159, 180

NAGPRA (Native American Graves Protection and Repatriation Act), 268
National Science Foundation (NSF), 66,

86, 169, 171, 222, 342, 343, 353, 401, 402
NATO conference in Chicago, 215, 216, 227
Negaunee Foundation, 353, 402
Nelson, Gareth (curator of ichthyology, American Museum of Natural History), 5, 7, 8, 12, 20, 363, 364
New Museum Idea, xiv
Nguyen, Thuy, 242, 243
Nitecki, Matthew (curator of invertebrate paleontology, Field Museum), 30, 33
Nordquist, Karen, 354
Novacek, Michael (curator of vertebrate paleontology and provost, American Museum of Natural History), 315, 316, 398
NSF. *See* National Science Foundation

Obama, Barack, 140, 274
Oceanview mine, 237-39, 245, 246

Pacifica hypothesis, 92-93, 101
Paleontological Institute of the Russian Academy of Sciences (PIN), 87-89, 102
Parenti, Lynne (curator of ichthyology, Smithsonian's National Museum of Natural History), 357
Parkinson, William (curator of anthropology, Field Museum), 174-76, 192, 193
patterns in nature. *See* evolutionary pattern vs. evolutionary process
Patterson, Bruce (curator of mammals, Field Museum), 286, 293-98, 302, 303, 396
Patterson, Bryan (curator of vertebrate paleontology, Field Museum), 291-93, 301
Patterson, Colin (curator of fossil fishes, Natural History Museum, London), 4, 5, 8, 11, 12, 14, 15, 361-63
Patterson, John, 285-92, 299, 396
peer community in science, 342, 343

Peru, 161, 173, 174, 183, 191, 273, 318, 333-37
Philippines, 306-13, 322-28
phylogenetics. *See* cladistics
Pritzker, J. N., 222, 228, 353

radioactive fossils, 89
Radke, Eric (Hans), xviii, 3, 35, 36, 47
rapid biological inventories, 317-20, 334-37
Ray, Clayton (curator of vertebrate paleontology, Smithsonian's National Museum of Natural History), 123, 130
rayfin fishes (definition), 86
Redden, David, 131, 133, 147
Ree, Rick (curator of botany, Field Museum), 159, 160, 180, 181
reef conservation, 313, 314, 329-31
religion vs. science, xi, xiii, 219, 343-45, 361, 362, 390, 402
repatriation, 268-73, 280, 281
research, basic (definition), 400
rhinos, 297, 298, 305
Rieppel, Olivier (curator of vertebrate paleontology, Field Museum), 28, 30, 32, 370, 371
Robbins, Helen (adjunct curator of anthropology, Field Museum), 269, 270, 272, 395
Rosen, Donn (curator of ichthyology, American Museum of Natural History), 5-7, 16, 17, 357, 364, 365
Rosenbaum, Tom, 391
Rowe, Jeanne, 353
Rowe, John, 353
Russia, 87-89, 102

Santucci, Vince, 121, 122, 129, 130, 145, 379
Schieffer, Kevin, 121-28
Schmidt, K-P (curator of herpetology, Field Museum), 196-209
science illiteracy, 342-47
science vs. religion, xi, xiii, 219, 343-45, 361, 362, 390, 402

Shaw, Bob, 354
Shaw, Charlene, 354
Shinya, Akiko, 63, 352
shrunken heads, 273–75
Shubin, Neil, 361, 391, 403
Siebert, Darrell (curator of ichthyology, Natural History Museum, London), 357
Sierwald, Petra (curator of myriapods and arachnids, Field Museum), 168, 168, 189, 382
Simms, Stephan (curator of anthropology, Field Museum), 266, 267, 277
Simpson, Bill, 132–34, 155, 379
Simpson, George Gaylord (curator of fossil mammals, American Museum of Natural History and Museum of Comparative Zoology, Harvard), 10, 12, 93
skeletons, cleared and stained, 6, 15, 17
Sloan, Robert, 3, 4, 38, 309
Society of Vertebrate Paleontology (SVP), 12, 119, 362, 363, 370, 378
Sotheby's auction house, 131–33, 147
special creationism. *See* creationism
specimen-based research. *See* collection-based research
Spielberg, Steven, 134, 135
Sterling, Eleanor, 316, 332, 398
Stones and Bones paleontology field course, 47, 62, 348, 352
Strong, William (curator of anthropology, Field Museum), 270
SUE the dinosaur, 110–55
Sytchevskaya, Eugenia (curator of fossil fishes, Paleontological Institute, Russian Academy of Science), 87, 88, 102

Tawani Foundation, 222, 353
tenure, 26, 27, 212, 223, 224, 368, 369, 383, 384, 402

Terrell, John (curator of anthropology, Field Museum), 176, 194
theory (in science), 359
Tsavo lions, 284–300, 302, 303, 396
Tynsky, Jim, 40, 49

Ulrich family, 37–39
UNAM, 65, 66, 74, 75

Vari, Richard (curator of ichthyology, Smithsonian's National Museum of Natural History), 357
Voight, Janet (curator of invertebrates, Field Museum), 167–68, 188, 382

Wadhwa, Meenakshi (curator of meteoritics, Field Museum), 163–65, 186, 382, 383
Wali, Alaka (curator of anthropology, Field Museum), 319, 320, 336, 398
Westneat, Mark (curator of ichthyology, Field Museum), 220, 390
Wiley, Ed (curator of ichthyology, Natural History Museum, University of Kansas), 357
Williams, Maurice, 116–18, 120, 128, 131
Williams, Ryan (curator of anthropology, Field Museum), 173, 174, 191, 280, 384
Wilson, E. O. (curator of entomology, Museum of Comparative Zoology, Harvard), 171, 217, 403
women curators (relative scarcity of), 382–84
Woods, Loren (curator of ichthyology, Field Museum), 363
Woods Hole Oceanographic Institution, 168, 188

Yabumoto, Yoshitaka (curator of paleontology, Kitakyushu Museum of Natural History), 92, 93